D1620068

Technology Transfer in Biotechnology

Edited by
Prabuddha Ganguli,
Rita Khanna, and
Ben Prickril

Related Titles

Lu, M., Jonsson, E., Speser, P. L. (eds.)

Financing Health Care

New Ideas for a Changing Society

2008

ISBN: 978-3-527-32027-1

Speser, P. L.

The Art and Science of Technology Transfer

2006

ISBN: 978-0-471-70727-1

Junghans, C., Levy, A.

Intellectual Property Management

A Guide for Scientists, Engineers, Financiers, and Managers

2006

ISBN: 978-3-527-31286-3

Budde, F., Felcht, U.-H., Frankemölle, H. (eds.)

Value Creation

Strategies for the Chemical Industry

2006

ISBN: 978-3-527-31266-5

Technology Transfer in Biotechnology

A Global Perspective

Edited by
Prabuddha Ganguli, Rita Khanna, and Ben Prickril

WILEY-VCH Verlag GmbH & Co. KGaA

The Editors

Dr. Prabuddha Ganguli
CEO, VISION-IPR, 101-201
Sunview Heights, Plot 262
Shere-e-Punjab
Andheri East, Mumbai 400101
India

Dr. Rita Khanna
International Technology
Transfer Management, Inc.
6533 Kenhill Road
Bethesda, MD 20817
USA

Dr. Ben Prickril
National Cancer Institute
Office of International Affairs
6130 Executive Boulevard, Suite 100
Rockville, MD 20852
USA

Library of Congress Card No.: applied for

British Library Cataloguing-in-Publication Data
A catalogue record for this book is available from the British Library

Bibliographic information published by the Deutsche Nationalbibliothek
The Deutsche Nationalbibliothek lists this publication in the Deutsche Nationalbibliografie; detailed bibliographic data are available on the Internet at 〈http://dnb.d-nb.de〉.

Printed in the Federal Republic of Germany
Printed on acid-free paper

Composition Asco Typesetters Ltd., Hong Kong
Printing Strauss GmbH, Mörlenbach
Bookbinding Litges & Dopf GmbH, Heppenheim
Cover Design Winkler

ISBN 978-3-527-31645-8

Contents

Technology Transfer in Biotechnology. A Global Perspective.
Edited by Prabuddha Ganguli, Rita Khanna, and Ben Prickril

ISBN: 978-3-527-31645-8

List of Contributors

G. Steven Burrill
Burrill & Company
Chief Executive Officer
One Embarcadero Center
San Francisco, CA 94111-3776
USA

Claudia Ines Chamas
Oswaldo Cruz Foundation
Ministry of Health
Av. Brasil 4365, Gomes Faria, sala 211
Rio de Janeiro, RJ 21045-900
Brasil

Marc Cornelissen
Bayer CropScience
Head of Research Operations–Bioscience
Technologiepark 38
9052 Gent
Belgium

Carlos Correa
University of Buenos Aires
Center for Interdisciplinary Studies of Industrial and Economic Law
Av. Figueroa Alcorta 2263
1425 Buenos Aires
Argentina

Prabuddha Ganguli
CEO, VISION-IPR
101-201 Sunview Heights
Plot 260 Sher-e-Punjab
Andheri East, Mumbai 400093
India

Florent Gros
Novartis Pharmaceuticals
25 Old Mill Road
Suffern, NY 10901-4106
USA

Gerald T. Keusch
Boston University
Associate Provost and Associate Dean for Global Health
715 Albany Street
Boston, MA 02118-2531
USA

Rita Khanna
International Technology Transfer Management, Inc.
6533 Kenhill Road
Bethesda, MD 20817
USA

Technology Transfer in Biotechnology. A Global Perspective.
Edited by Prabuddha Ganguli, Rita Khanna, and Ben Prickril

ISBN: 978-3-527-31645-8

Lita Nelsen
Massachusetts Institute of Technology
Technology Licensing Office
Five Cambridge Center Kendall Square
Cambridge, MA 02142-1493
USA

Yukiko Nishimura
University of Tokyo
Research Center for Advanced Science and Technology
4-6-1 Komaba, Meguro-ku
Tokyo 153-8904
Japan

Ben Prickril
KenGar Consulting
14, Rue Lavoisier
69003 Lyon
France

Mark L. Rohrbaugh
Currently
The REDANDA Group, Inc.
PO Box 305
Monrovia, MD 21770
USA

Formerly
National Institutes of Health
Office of Technology Transfer
6011 Executive Boulevard
Rockville, MD 20852
USA

Brian R. Stanton
Currently
The REDANDA Group, Inc.
PO Box 305
Monrovia, MD 21770
USA

Formerly
National Institutes of Health
Office of Technology Transfer
6011 Executive Boulevard
Rockville, MD 20852
USA

Ashley Steven
Boston University
Office of Technology Development
53 Bay State Road
Boston, MA 02115
USA

Katsuya Tamai
University of Tokyo
Research Center for Advanced Science and Technology
4-6-1 Komaba, Meguro-ku
Tokyo 153-8904
Japan

Michiel M. van Lookeren Campagne
Bayer CropScience
Head of Bioscience Research
Technologiepark 38
9052 Gent
Belgium

Jianyang Yu
Liu, Shen & Associates Law Firm
Hanhai Plaza, 10th Floor
10 Caihefang Road
Beijing 10080
China

1 Defining the Future: Emerging Issues in Biotechnology, Intellectual Property Rights and Technology Transfer

Prabuddha Ganguli, Rita Khanna, and Ben Prickril

1.1 Introduction

Since the formation in 1976 of the first modern biotechnology company, Genentech, the biotechnology industry has grown to become one of the major engines of innovation in virtually all developed economies. Indeed, biotechnology's growth in areas ranging from health, agriculture, environment and industrial processes has been phenomenal. This expansion has been paralleled by mounting public concerns because of potential ethical issues and impact on our health, food and the environment.

The importance of innovation in biotechnology and its widespread applications in health, agriculture and commerce has helped bring issues related to intellectual property (IP) rights and technology transfer into sharp focus. The ongoing global debate on IP rights, especially related to health and agriculture, has hinged on proprietorship of knowledge and its ethical and political implications for innovation, knowledge sharing and technology transfer. The means by which knowledge and technologies are moved from basic research up the value chain to become commercial products is critical to the ability of biotechnological innovation to reach those who need it.

1.2 Historical Evolution of Intellectual Property Regime in Biotechnology

There has been a marked paradigm shift in the field of IP rights itself, especially in the areas of patents and copyrights. The modern patent system originated in 1474 as a means of providing inventors the right to block others from using their inventions in return for registering them with the government.

Early inventions usually dealt with the creation of inanimate and tangible objects, but as understanding of basic phenomena progressed, inventions relating to intangibles became fairly common. The field of biotechnology IP rights is

Technology Transfer in Biotechnology. A Global Perspective.
Edited by Prabuddha Ganguli, Rita Khanna, and Ben Prickril

ISBN: 978-3-527-31645-8

often intangible and became even more so when the field was transformed by the advent of molecular biology in the 1960s and 1970s. The tools of molecular biology began to enable production of completely new therapeutic drugs, vaccines, diagnostic tools and plant breeding methods starting at the level of individual genes. The seminal US Supreme Court Case *Diamond v. Chakrabarty* was the turning point in the history of IP rights related to biotechnology. Since the Supreme Court's ruling in *Diamond v. Chakrabarty*, certain non-naturally occurring organisms are eligible for patent protection and the patent system has played a critical role in stimulating an emerging biotechnology industry. This decision led the way to patenting life forms provided they were created by human intervention, and met the requisite criteria of novelty, inventive step and utility. Patent exclusivity for biotechnology inventions catalyzed further investments in R&D in biotechnology and marked the dawn of a new biotechnology industry. The tremendous development of this industry and the concomitant increase in the proprietorship of knowledge through IP rights has raised contentious issues in knowledge transactions in a competitive environment.

1.3 Issue of Patentability of Gene Sequences, Antibodies, Early-Stage Technology/ Platform and 'Insufficient Support for Claims'

The growth of biotechnology has presented new challenges to the patent system. As noted above, right from the outset issues of what is patentable and how it should be patented have been particularly important and contentious in the biotechnology field. Some aspects have been clarified and resolved, while others still remain to be addressed and new issues continue to emerge. This section will review the history and remaining issues with respect to the patentability of genes, antibodies, research tools and platform technologies.

A subject in biotechnology that has attracted critical attention is the subject matter of erythropoietin. A 2004 UK House of Lords Decision invalidating Kirin–Amgen's erythropoietin patent questioned the patentability of gene sequences as the court observed that 'gene sequences are to be assessed as "discoveries" or just "information about the natural world"'. This decision suggests that the bar to patentability in matters related to gene sequences needs to be regularly reassessed.[1)]

The patentability of antibodies has also been questioned in several recent decisions as in *Noelle v. Lederman* [355 F.3d 1343, 1349 (Fed. Cir. 2004)] and *Smithkline Beecham v. Apotex* [403 F.3d 1328 (Fed. Cir. 2005)]. In the case of *Noelle v. Lederman*, the court observed that the written description of the specification did not provide sufficient support for claims to a human antibody because it failed to disclose the structural elements of the human antibody or antigen. In the *Smith-*

1) Crespi, R.S. (2005) Erythropoietin in the UK: a setback for gene patents? *Nature Biotechnology*, **23**, 367–8.

kline Beecham v. Apotex matter, the court observed that there was 'inherent anticipation'.[2]

The ongoing case of *Novartis Pharmaceuticals Corp. v. Teva Pharmaceuticals USA, Inc.* [05-1887, 2007 WL 2669338 (DNJ 6 September 2007)][3], addressing issues related to obviousness, provides insight regarding the importance of references that teach away from an invention. The question being addressed is whether a specific article teaches away from penciclovir, but the prior art 'as a whole' did not teach away from using penciclovir as a lead compound.

Patenting of early-stage technologies such as target identification, pathway analysis, platform technology development and even generation of putative biotherapeutic compound leads have also been subject to debate. In several cases that have come before the courts especially in the United States, such as *The University of California v. Eli Lilly* [119 F.3d 1559, 43 USPQ2d (BNA) Fed. Cir. 1997), cert. denied 523 US1089 (1998)], *Amgen v. Chugai* [927 F.2d 1200, 18 USPQ2d (BNA) 1016 (Fed. Cir. 1991)] and *Fiers v. Revel* [984 F.2d 1164, 25 USPQ2d (BNA) 1061 (Fed. Cir. 1993)], it has been clearly shown that when such patents are challenged, they have not stood the test of validity regarding an adequate written description of the invention.[4]

The field of genomic diagnostics and IP rights is also becoming embroiled in controversy. Most debates have centered around the patents on *BRCA1* and *BRCA2*, questioning the intent of the patent holders to unreasonably restrict access to the important diagnostic tests. In an article titled 'Emerging patent issues in genomic diagnostics',[5] Barton raises several questions especially on the problem of royalty stacking. There could be a series of patents claiming the use of a specific gene sequence to identify a specific biological property that may make it difficult for the integrator of a microarray/chip device to assemble the rights to use the different patented sequences that are relevant to the clinical or research application. In principle, each holder of a patent on a diagnostic sequence marker used in the array could traditionally block marketing or the use of the array. Similarly, patents may be issued on sequences that might identify drug efficacy or side-effects. Such patents may cover sequences as biomarkers of an effect on drug metabolism, or the use of sequences to make decisions about drug regimes. Barton suggests that the patent law needs to be assessed keeping in mind such developments in the field of biotechnology and to improve access to the pool of available knowledge.

Another issue that is gaining prominence is the question of 'patenting race'. An article by Khan[6] raises issues related to the strategic use of race as a genetic category to obtain patent protection and drug approval as they are increasingly

2) Lu, D.L., Collinson, A.M. and Kowalski, T.J. (2005) The patentability of antibodies in the United States, *Nature Biotechnology*, **23**, 1079–80.

3) Lu, D.L., Collission, A.M. and Kowalski, T.J. (2007) Patentability issues surrounding antivirals, *Nature Biotechnology*, **25**, 1403–4.

4) Suster, M.J., Su, H. and Blaug, S. (2003) Protecting rights to early-stage technology, *Nature Biotechnology*, **21**, 701–3.

5) Barton, J.H. (2006) Emerging patent issues in genomic diagnostics, *Nature Biotechnology*, **24**, 939–1.

6) Kahn, J. (2006) Patenting race, *Nature Biotechnology*, **24**, 1349–51.

being evoked in biotechnology patents. Between 1976 and 1977 there were no issued patents in the United States that mentioned racial and ethnic categories. However, during the period 1998–2005, there were a total of 12 instances in issued patents in which race and ethnic categories were mentioned. Further, in patent applications from 2001 to 2006, there were 65 instances in which race and ethnic categories were mentioned. In June 2005, BiDil became the first drug approved by the US Food and Drug Administration (FDA) with a race-specific indication. Underlying BiDil's New Drug Application for FDA approval is a 2002 race-specific patent specifying use of the drug for treatment of heart failure in an African-American patient (US 6465463). Interestingly NitroMed, BiDil's corporate sponsor, also holds an earlier patent (US 4868179) to use BiDil in a general population, regardless of race. The earlier patent expired in 2007, whereas the race specific patent expires in 2020.

Similarly in Europe, in June 2005, the European Patent Office upheld a patent owned by Myraid Genetics relating to the testing for *BRCA2* genetic mutation for 'diagnosing a predisposition to breast cancer in Ashkenazi Jewish Women'.[7] Such patents will have profound sociological and economic consequences in due course.

1.4 Scope of Patent Claims

As in other areas of IP rights, the appropriate term, breadth and specificity of patents has been a continuing and, indeed, a growing concern. The number of patents issued has grown exponentially. Proponents of more stringent IP rights have stressed the importance of a robust system of patents for biotechnology. On the other hand, many have raised concerns about the excessive number and breadth of patents, and their growing complexity of knowledge sequestration is discouraging efficient diffusion of knowledge and undermining research. Striking a balance between adequate IP rights protection and the efficient availability of knowledge with spillover effects remains a continuing challenge.

Patenting of research tools has been at the center of an important debate over the last decade, without much clarity of date. These tools are generally recognized as embracing the full range of resources that scientists use in the laboratory, including such items as cell lines, animal models and reagents.[8]

Several areas such as patenting of expressed sequence tags (ESTs), which are essentially research tools, have been perceived to severely restrict research, while being unlikely to result in discrete commercial products. One means of addressing these concerns has been to raise the bar to utility, as was done through the

7) Kienzien, G. (2005) *The Scientist*, **July 1**, http://www.the-scientist.com/article/display/22719.

8) NIH (1998) *Report of the NIH Working Group on Research Tools*, June 4, NIH, Bethesda, MD. Available at http://www.nih.gov/news/researchtools/.

Utility Examination Guidelines set forth by the US Patent and Trademark Office (PTO) in January 2001. The case in question that issues in EST patenting is well discussed in an article by Davis *et al.*[9] The article elaborates on Ficher's claims in which Monsanto scientists in their patent application disclosed approximately 32 000 specific nucleotide sequences for ESTs of various maize tissues. Although, during the patent prosecution, the PTO restricted Monsanto to five ESTs, the patent claim read 'A substantially purified nucleic acid molecule that encodes a maize protein or fragment thereof comprising a nucleic acid sequence selected from the group consisting of SEQ ID NO:1 through SEQ ID NO:5'. Such a claim effectively covers any purified nucleic acid that includes one of the five ESTs so long as the nucleic acid (not necessarily its EST portion) encodes a maize protein or even a fragment of a maize protein. The question raised with such broadly granted claims is whether they would prevent basic genomic research or deny the use of associated proteins as targets for product screening.

'Reach-through' claims to drug targets have a major impact on ownership and, therefore, control on future activities involving drug targets, and have been the subject of much debate. In several cases, if a variant has less side-effects or is more effective, the first patent with 'reach-through' claims could preclude the development of drugs for the variant target forms. Under such circumstances, there are varying opinions on the options that can be exercised, one of them being that the holder of patent rights has a strong incentive to negotiate licenses to subsequent drug developers or to variations in the metabolic pathways that breakdown the drug.[10]

Patenting in areas related to stem cells, especially in terms of claiming proprietary rights to 'pluripotency', is opening up new challenges to the drafting of patent specifications in terms of what would be considered as adequate disclosures, allowable and enforceable claims construction, and examination processes in patent offices and further the framing of national policies.

Recent proceedings in the PTO *vis-à-vis* the rejection, re-examination and allowance of patents especially related to US 5843780 (claiming pluripotent primate embryonic stem cells and a method of isolating a primate embryonic stem cell line), US 6200806 (claiming pluripotent human embryonic stem cells and a method of isolating a human embryonic stem cell line), US 7029913 (claiming pluripotent human embryonic stem cells) and rejection of the continuation application US 20050158854 (claiming pluripotent human stem cells) are of relevance as they provide directional indications on future approaches likely to be taken by patent offices, especially in the United States, to arrive at conclusions on 'obviousness' and 'enablement requirements'.[11] The key questions that have

9) Davis, P.K., Kelley, J.J., Caltrider, S.P. and Heining, S.J. (2005) ESTs stumble at the utility threshold, *Nature Biotechnology*, **23**, 1227–9.

10) Bohrer, R.A. (2008) Reach-through claims for drug target patents: Rx for pharmaceutical policy, *Nature Biotechnology*, **26**, 55–6.

11) Vrtovec, K.T. and Scott, C.T. (2008) Patenting pluripotence: the next battle for stem cell intellectual property, *Nature Biotechnology*, **26**, 393–5.

surfaced with regard to the granted patents US 6200806 and US 7029913 at the re-examination stage relate to 'whether methods described in the prior art would extend to the method of isolation of embryonic stem cells?' and 'whether the techniques used were unpredictable and not universally applicable to the isolation of embryonic stem cells from other species, particularly human'. Similarly, while rejecting the claims of the continuation application US 20050158854, the argument by the PTO has been 'because the specification, although being enabling for the preparation of pluripotent hES [human embryonic stem] cells, does not reasonably provide enablement for a preparation of pluripotent hES cell'. Further the question that needs resolution is whether the term 'pluripotent hES cells' cover 'human iPS [induced pluripotent stem] cells'. The answer to this question would be based on the definition of these terms which hopefully will get clarified by courts in future.

The matter is of immense interest in view of the patent application by Yamanaka (WO/2007/069666 published on 21 June 2007). This patent discloses a means for inducing the reprogramming of a differentiated cell without using any embryo or embryonic stem cell, thereby establishing an inducible pluripotent stem cell having similar pluripotency and growing ability to those of an embryonic stem cell with good reproductivity. This is accomplished by a nuclear reprogramming factor for a somatic cell comprising products of the following three genes: an *Oct* family gene, a *Klf* family gene and an *Myc* family gene. The Japanese Patent Office has recently granted the patent 2008-131577 in Japan, keeping a lead in the induced pluripotent stem cell patent race.[12)]

Two recent examples illustrate tensions among various stakeholders involved in the commercialization of such developed technologies.

Monsanto Technology LLC v. Cargill International SA and Cargill PLC [Neutral Citation Number: [2007] EWHC 2257 (Pat) Case No: HC06C00585; http://www.bailii.org/ew/cases/EWHC/Patents/2007/2257.html], a litigation in the United Kingdom, serves as an example of how construction of DNA-based claims could be interpreted in terms of the specificity or breadth of the claims. Monsanto sued Cargill for importing into the United Kingdom soya meal produced in Argentina alleged to be derived from soya beans modified to contain a gene conferring resistance to a herbicide called glyphosate (Roundup). Cargill counterclaimed for invalidity of the Monsanto patent and also contested infringement. The judge found the patent valid as amended by Monsanto, but not infringed by Cargill's importation of the soya meal.

Claim 1 of the EP (UK) 0546090 states:

An *isolated* DNA sequence encoding a Class II EPSPS [5-enolpyruvylshikimate-3-phosphate synthase] enzyme, said enzyme being an EPSPS enzyme having a K_m for phosphoenylpyruvate (PEP) between 1–150 μM and a K_i(glyphosate)/K_m(PEP) ratio between 3–500, which enzyme is capable of reacting with antibodies raised

12) Cyranoski, D. (2008) Japan fast-bracks stem-cell patent, *Nature*, **455**, 269.

against a class II EPSPS enzyme selected from the group consisting of the enzymes of SEQ ID NO: 3 and SEQ ID NO: 5, *which DNA sequence encodes the amino acid sequence of SEQ ID NO: 3 save that serine at position 2 is replaced by leucine.*

The judge agreed with the interpretation that the term 'isolated' when read in the context of the invention disclosed in the patent described the creation of transgenic plants that were resistant to glyphosate and in order to achieve that aim, the gene sequences in question should be purified and available for manipulation for use in the processes claimed in the patent. The judge found that Monsanto had not (i) completed experiments to demonstrate that all of the test criteria set out in Cargill's construction had been met, or (ii) proved that the DNA that was actually present in the soya meal was 'isolated', thereby resulting in the finding that Cargill did not infringe the DNA product claims of the said patent.[13)]

This decision highlights the manner in which patent claims may be interpreted narrowly without allowing broad interpretation of terms used in the claims, and further suggesting that claims to the production of a genetically modified organism may not be 'proximal' enough to derivative products to be deployed to prevent their importation.

It has now been well documented that there has been gradual decline in the filing and granting of patents claiming DNA sequences. However, patent offices have been granting patents claiming DNA with narrower scope and relatively robust claims especially claiming splice variants and single nucleotide polymorphisms.[14)]

1.5 Institutional Arrangements for Technology Transfer

A policy concern that arose early in the drug development field was how to create adequate incentives for commercialization of results of basic research. In a system where basic research was primarily being undertaken by academic and public institutions, there was a concern that this research was neither being utilized nor producing adequate returns for the taxpayer. One important manifestation of this concern relates to the lack of incentives for the development of medicines for developing country diseases. Although most governments have instituted departments to fund and administer R&D the issue of IP rights and technology transfer between institutions remains a major bottleneck. New mechanisms (i.e. public–private partnerships) is to develop and manage IP are beginning to have a positive impact for a large number of people in developing countries to meet their need for better access to food and medicines.

13) Cohen, S. and Morgan, G. (2008) Monsanto Technology LLC v. Cargill: a matter of construction, *Nature Biotechnology*, **26**, 289–91.

14) Hopkins, M.M., Mahdi, S., Patel, P. and Thomas, S.M. (2007) DNA patenting: the end of an era? *Nature Biotechnology*, **25**, 185–7.

An example of intra-institutional collaboration is the establishment of the Oxford Genetics Knowledge Park (OGKP) in 2002, which was a partnership between Oxford University and the Oxford Radcliffe Hospitals NHS Trust funded by the UK Department of Health/Department of Trade and Industry with the aim of translating advances in genetics research into clinical practice. Important issues identified by the OGKP for effective working are research exemptions and their applicability, landscaping of patents to minimize risk of infringement, and need for infrastructure for support.[15] There are several examples of technology development and transfer by and between institutions.[16]

Easy access to proprietary information is of significance to researchers especially in universities to reduce their risks for patent infringement. Several resources such as the Public Intellectual Property Resource for Agriculture (www.pipra.org), CAMBIA (www.cambia.org) and several other organizations are focusing their efforts to promote an open forum to assist in IP rights-related matters especially for researchers in universities.[17]

The 'Cooperative Research and Technology Enhancement Act of 2004' (CREATE ACT) became effective in September 2005 in the United States, and is intended to promote collaboration among industry collaborators, and therefore promote innovation, decrease costs, and ultimately enable the commercialization of patented biotech products and processes for societal good. This Act will have major implications on technology transfer and IP rights in areas related to biotechnology. However, to make the CREATE ACT meaningful, it will be important to address issues related to 'research exemptions' and diagnostic tools, and to ensure freedom of researchers to use patentable inventions for their research.[18]

1.6 Policy Issues and Challenges

Technology transfer in biotechnology depends on the transformation of basic research findings into commercial products and requires a strong IP rights system to succeed. Along the path, there are a host of issues that need to be addressed. These include:

- Knowledge 'sequestration' caused by proliferation of biotechnology patents, thereby placing knowledge into privatized 'knowledge black holes' and patent thickets, therefore making it accessible to others.

15) Kate, P., Hawkins, N. and Taylor, J. (2007) Patents and translational research in genomics. *Nature Biotechnology*, **25**, 739–41.

16) Ganguli, P. (Guest ed.) (2005) Special volume 'Technology Transfer with IPR', *Journal of Intellectual Property Rights*, **10**, 349–456.

17) Yancy, A. and Stewart, C.N. Jr (2007) Are university researchers at risk for patent infringement? *Nature Biotechnology*, **25**, 1225–8.

18) Mills, A.E., Chen, D.T., Gillon, J.J. Jr and Tereskerz, P.M. (2006) The CREATE Act: increasing costs associated with the biotech industry? *Nature Biotechnology*, **24**, 785–6.

- Instances of abusive monopoly resulting in higher prices for patented medicines and other products.
- Granting of broad claims by various patent offices leading to excessive patent protection.
- Freedom to operate restrictions on academic researchers due to patenting of research tools and issues related to non-clarity on 'research exemptions' undermining scientific progress.
- Increasing complexity of licensing deals resulting in increased research and transaction costs, including litigation.
- Privatization of patents from government-funded R&D by universities and research institutions, especially where little benefit accrues to the general public.

Countering these concerns, proponents of a strong IP rights regime argue that without adequate IP rights protection, transfer of technologies would be stymied, and investments in R&D would not yield meaningful returns due to negative impacts on both the potential for further investments and on innovation and knowledge diffusion to the detriment of society. Proponents also suggest that a strong IP rights system would promote effective and appropriate technology transfer and the development of human resources in lesser-developed countries lacking the critical mass of expertise and infrastructure in biotechnology.

Frameworks of IP rights clearly need to keep pace with the rapid developments in science and technology in order to create an enabling platform for legitimate access to information. Sharing and using knowledge equitably benefits all stake holders and promotes the sustainable growth of the global society. In the same way, IP rights issues that impact on human rights or raise ethical concerns need to be analyzed systematically and take into account the concerns in national and international policy affecting technology transfer in biotechnology.

An Organization for Economic Cooperation and Development (OECD) report[19] exhaustively addresses the issues related to genetic inventions, IP rights and licensing practices. Some of the key conclusions of the OECD report are:

- *Responsibility of the patent offices.* The scope of patent claims in relation to biological and genetic material continues to be a matter of concern. For example, broad claims in several cases have been allowed by patent offices so as to give rights to the patent applicant on the genetic makeup of plants and organisms beyond individual varieties, species and genera to incorporate key elements of genomes across classes, which in effect cover species that have not been invented or even known. Patent offices should, therefore, issue clear guidelines on benchmarks for patentability ('raising the bar') so that examiners do not issue patents wrongly or allow exceptionally broad claims.
- *Need for public engagement.* Enhancing awareness and engaging the public in debates on applications of biotechnology, patenting, technology transfer, etc.,

19) Organization for Economic Cooperation and Development (2002) *OECD Report of the Workshop on Genetic Inventions, Intellectual Property Rights and Licencing Practices, 24 and 25 January*, OECD, Berlin, Germany. Available at http://www.oecd.org/dataoecd/7/42/194903.pdf.

will be crucial if public trust in the patent system and its application to biotechnology is to be strengthened.
- *Institutional arrangements for the promotion of technology transfer.* Policies dealing with biotechnology, IP rights and technology transfer in academia and industry need to be aligned that academia using public funding is not hindered in making use of patented inventions without liability. In this regard, governments should harmonize their policies on the subject of 'research exemptions' so that there is some uniformity in interpretations in diverse jurisdictions.
- *Governmental policies.* Objective monitoring of patenting and licensing of genetic inventions should become the basis for policy making to ensure fair and reasonable access to genetic information and subject matter of patented inventions.
- *Anti-competitive practices.* Guidelines for the use of provisions by governments on issues related to anti-competitive practices in contractual licenses, effective use of competition law to control abusive exploitation of IP rights, benchmarks for the issuance of compulsory licenses need to be defined more clearly so that IP rights are used with fair benefit sharing between the diverse stake holders in the interest of societal growth.

It must be recognized that the patent system was set up to stimulate innovations. The idea was that a patent would give the inventor rights to their invention in return for disclosing it so that others who would have access to the knowledge would be able to invent around or contribute through improvements. However, in the case of biotechnology, and especially in systems biotechnology, allowing inventions related to basic biological processes often leaves no options for other inventors to invent around, thereby giving a virtual monopoly to the patent holder of the basic patent. This raises serious questions about the granting of patents for basic processes in biology, which seemingly conflicts with the very purpose of granting of patents. There is a growing feeling that IP rights in biotechnology are possibly indirectly denying the public at large some of the biomedical and agricultural benefits that they rightly deserve as a part of the social contract between the inventors and the society.

The Bayh–Dole Act in the United States allowed universities to own their government funded inventions and license them to commercial partners. However, universities are increasingly under pressure to sign contracts with restrictive non-disclosure agreements for their privately funded research, causing undue delays in the timely sharing of their findings with their peer groups. A movement of 'Creative Commons' is also gaining momentum with an increasing concern on establishing a research exemption from infringement of gene-related patents. With lessons learnt from the experiences of implementing the Bayh–Dole Act in the United States, similar legislations are under active consideration in several countries.

It has been reported that 20% of the human genome is claimed as patents, of which two-thirds are owned by private firms. Some of these patents appear to have been granted with broad claims that are in themselves questionable due to the limited ability of patentees to satisfy the utility (industrial applicability) requirement for their invention. The creation of strategic patent estates by private firms using such patents is now being seriously questioned on the basis that they are leading to a virtual monopoly and underuse of the developed knowledge for the social good.

Further, as genetic testing moves into mainstream medicine, the effect of gene patents will have far-reaching effects on the healthcare system and the industry servicing it, and therefore will have to be handled with care.[20]

Patent filing in the area of stem cells is on a steady growth path, and recent patents granted to Wisconsin Research Foundation and Kyoto University research groups have become very controversial.[21] The emergence of patent thickets is likely to cause problems in 'freedom to operate', imposing multiple layers of transaction costs and stacking of royalty payments beyond levels that can be supported by the value of single innovations. Fears have been expressed about the possibility of slower movements of innovations to the market place, dampening investor of confidence, and the undermining the transfer of technology and networking to promote public–private partnerships. The need to establish institutionalized models for collective networking involving 'technology development conglomerates', 'consortia technology development programmes with appropriate process for IP ownerships and equitable benefit sharing among participating partners', open-source licensing, formation of patent pools and IP rights warehousing including clearing mechanisms will pave the way for the development of technologies, making their benefits accessible to society in a cost-effective and affordable manner. Caution is to be exercised while performing in such collective modes in order not to form coercive groups and promote anti-competitive frameworks.

The impact of some recent US Supreme Court decisions on license negotiations, finality of licensing agreements, and control by licensors can exert a detrimental impact on the downstream use of their licensed technologies and, therefore needs to be appreciated with special reference to cases such as *eBay, Inc. v. MercExchange LLC*, *MedImmune, Inc. v. Genentech, Inc.* and *Quanta Computer, Inc. v. LG Electronics* as they will begin to shape the cooperative trajectories of the future.[22]

The convergence of nanotechnology, biotechnology, information technologies and microelectronics has opened up new opportunities in areas as diverse as drug delivery systems, tailored tissue engineering, microfabrication, biosensing devices and microarray technology. The development of these new functions will require a high level of expertise and sophistication in the management and transfer of technology. Both developed and developing countries need to ensure that their IP regimes and technology transfer mechanisms are able to keep pace with these developments.

20) Klein, R.D. (2007) Gene patents and genetic testing in the United States, *Nature Biotechnology*, **25**, 989–90.

21) Bergman, K. and Graff, G.D. (2007) The global stem cell patent landscape: implications for efficient technology transfer and commercial development, *Nature Biotechnology*, **25**, 419–24.

22) Giordano-Coltart, J. and Calkins, C.W. (2008) Recent Supreme Court decisions and licencing power, *Nature Biotechnology*, **26**, 183–5.

Part I
Technology Transfer Policy Considerations and Country/Regional

Technology Transfer in Biotechnology. A Global Perspective.
Edited by Prabuddha Ganguli, Rita Khanna, and Ben Prickril

ISBN: 978-3-527-31645-8

2
Technology Transfers in Europe within the Life Sciences

Jacques Warcoin

2.1
Biology and the Development of Technology Transfers

Technology transfers within the life science sector have developed exponentially since the advent of biotechnology in the 1980s. There are many reasons for this, but the main one is related to the particular structure of this emerging sector.

The first patent in the biotechnology sector (US Patent 4,237,224, granted on 2 December 1980) was granted to Cohen and Boyer (Nobel Prize for Biology) who discovered the tools of biotechnology. Thanks to the young and enthusiastic venture capitalist Robert A. Swanson, who persuaded Boyer to create a company to exploit his invention, Genentech, the first biotechnology start-up, was created.

At the cost of a number of contacts with major pharmaceutical companies, the creators of Genentech noted that these 'big pharma' were not directly interested in developing biotechnology in-house. In actual fact, a large number of companies contacted were interested in the principle, but not in the facts – those companies taking the view that the technology was too advanced and they therefore did not want to take direct risks in this type of development.

Genentech, like other start-up companies that had just been formed, realized very quickly that, in the pharmaceutical sector, a small company had practically no chance of bringing a product to the marketing stage. It was therefore necessary to be able to develop the company in spite of it being impossible to market products, at least to start with.

This was why the biotechnology start-up companies took the view, from the outset, that only industrial property could be rapidly developed, and this gave the impetus to an intensive and aggressive industrial property policy, particularly the filing of patents, but also regarding contractual negotiations that had to take into account the type of companies involved – in most cases a very small company facing major pharmaceutical groups.

Even if such a situation was obviously not new, it became more or less the norm and therefore involved the creation of specific rules. The first result

Technology Transfer in Biotechnology. A Global Perspective.
Edited by Prabuddha Ganguli, Rita Khanna, and Ben Prickril

ISBN: 978-3-527-31645-8

therefore was a proliferation of R&D contracts and license contracts between the start-up companies and the major pharmaceutical groups.

2.2
Involvement of Public Bodies

In many countries, and particularly in Europe, research in the biology field is carried out largely within public bodies (e.g. universities and institutes). This has led to the creation of start-up companies that are in fact spin-offs from university laboratories but which, for that very reason, needed to develop complex contractual networks between the public bodies and these new private companies.

These contracts are complicated by the special status of certain university researchers who are civil servants of European countries and who, for that reason, are not always at liberty to participate in the formation of a start-up company without abandoning their civil servant status (e.g. in France).

A second feature of this field is that these contracts between public bodies and private companies meet obligations that differ greatly according to the country.

Finally, beyond the 'major company–start-up–public body' triangle, the final feature of this field is the development of very sophisticated industrial property strategies, in particular 'meshes' of patents and various rights, such as plant variety rights, database law and the protection of know-how, which have progressively led to industrial property situations that could make it impossible to carry out certain projects ('patent thickets' and freedom to operate).

Taking their inspiration from solutions to be found in sectors like electronics, the life sciences have extracted certain solutions and transposed them into their sector. However, these solutions are not apparently transposable as such and, in that case also, adaptations are necessary to enable these contractual technologies to be used for technology transfers in general. This applies, for example, to adaptations made for patent pools and clearing houses that are already used in the electronics sector and new information technologies.

Ultimately, as was the case in the electronics sector, the situation became so excessive in some cases that it led to the appearance of what are known as patent trolls – companies that are technology transfer companies that merely use the system to collect very high royalties – which constitute real obstacles to the development of projects and which, in that case also, require specific strategies to be applied, both in terms of contracts and, possibly, in terms of behavior during negotiations.

The purpose of this chapter is therefore to present a number of clauses that have now become 'conventional' in the field of negotiations in the life sciences, clauses which until now were not used in this form or possibly some forgotten clauses that have taken on a new aspect and a new significance in a sector as complex as this.

It is of course impossible, when dealing with the situation of civil servants with regard to their supervising bodies, to list the situations in the various countries of the European union and for this reason it will in some cases simply be stated that this point must be specifically considered when precontractual contacts are made.

2.3 Initial Contacts

In the biotechnology sector mistrust is often the rule, considering the disproportions between the various players and often the differences in their status (start-up as opposed to major company, public as opposed to private). Mistrust leads to almost as many mistakes as reasonable trust, but above all it enormously prolongs the negotiation period.

For this reason, 'attentive' trust is preferable and, as has been suggested a number of times, one must always start by 'reading on the lines' before 'reading between the lines'. An American author said that this was less tiring for the eyes, but is also a way to be sure that one has at least properly understood the essence of the contract.

Nevertheless, the initial contacts are always laborious. In the case of a start-up and major public bodies, or indeed contact with major pharmaceutical companies, one can be certain that the start-up companies also very frequently take part in their first negotiations represented by scientists whose knowledge with regard to contractual negotiations is at least quite limited, not to say frequently very mistaken.

Under these circumstances, the negotiations often take place with people who have heard that 'one of their friends was plundered by...' or else, on the contrary, between people who respect each other at a scientific level, but who totally misunderstand the elementary rules that must govern initial contacts in terms of technology transfer. This is why one frequently faces pre-existing situations that may become insoluble simply because the initial contacts were made in circumstances that are deplorable in legal terms.

The first recommendation is therefore to warn any scientists making contact for the purposes of any negotiation whatsoever that they must not make any commitment without the agreement of the legal managers and that they absolutely must be covered by at least one confidentiality agreement or non-disclosure agreement.

This comment may appear commonplace, but in the biotechnology sector it appears that, in the case of start-ups, almost all initial contacts with a view to a technology transfer are made by scientists who ignore these elementary precautions.

2.4 Non-Disclosure Agreements

The non-disclosure agreement (NDA), secrecy agreement or equivalent is generally a preliminary to the first precontractual exchanges. In the biotechnology sector, more than in any other sector, NDAs must contain annexes that very precisely define the information covered by the NDA. This is because most of the time the relevant companies are working on the same subject, and it is therefore necessary for each of them to be able to determine very precisely what must be covered by confidentiality and what in actual fact already belongs to the other party and is not covered by confidentiality.

Furthermore, in this sector where one is frequently dealing with scientists, it is necessary to know whether the scientist who is to sign the NDA is properly authorized to bind his company and that he is not signing in a personal capacity.

Although some countries recognize the concept of 'apparent authority', meaning that the signatory presents themself and has all the apparent characteristics to be able to bind their company, other countries on the contrary consider that no 'apparent authority' exists and that only a written authorization can bind the company.

Particular attention must be paid to the content of this confidentiality agreement, which sometimes includes clauses that do not relate to the confidentiality obligation as such but, for example, to the intellectual property (IP) of the results of the collaboration being contemplated – clauses that should be avoided as they are premature at this stage.

The confidentiality obligation itself may be limited to information within a given scientific field or it may be extended to all information exchanged between the parties, regardless of the subject with which they are dealing. The confidential nature of this information may be determined *a priori* (information is confidential as soon as it is communicated and comes within the scope of the contract) or *a posteriori* (the information only becomes confidential when it is confirmed as such, e.g. by a 'confidential' endorsement on the document containing it or by written confirmation of a telephone conversation).

A point that is generally not often dealt with in this type of agreement relates to the obligation to communicate information. Is a party that holds information that may come within the sector of the contractual negotiations obliged to communicate it? In the absence of any provision, the French courts will apply the principle of good faith in the performance of contracts. They will examine, in particular, whether the failure to communicate the information in question does or does not constitute an act of bad faith characterizing the presence or absence of a breach of contract. A clause that makes such communication obligatory will mean, if it is violated, that it is unnecessary to prove bad faith to the court, as the mere failure to communicate in itself constitutes the breach of contract.

The question of the use which will be made of the transferred information is also of major importance. Indeed, the use which will be made of the communicated information must be precisely defined, which is a sometimes neglected corollary of confidentiality.

It is recommended that provision be made for this use to be strictly limited to the purposes of the precontractual negotiations, in particular to the assessment of the prospects offered by the relevant technology. Reverse engineering, or the analysis of samples, must be expressly excluded unless the communicating party has given its prior consent and within the limits of the right of use.

The end of the contract is often neglected (and confidentiality contracts are not alone in that respect), whether it be the question of the expiry date of the contract, what happens to information communicated subject to confidentiality or else the outcome of the precontractual negotiations.

The expiry date of the contract, which is generally set at the expiration of a certain period, does not bring to an end the need to maintain the obligations communicated under the confidentiality system. This is because it must not happen that, the day following the end of the negotiations, regardless of the outcome thereof, the beneficiaries of this information be able to disclose them to third parties with total impunity. This is why it is preferable to make provision that, at the end of the contract, the confidentiality obligation 'survives' this contract, for a period generally between 1 and 5 years, more rarely 10 years, depending on the strategic nature of the relevant information.

In any event, the end of the contract must provide for the return or destruction of the documents embodying the confidential information. The parties sometimes provide that a copy of the documents may be retained by the beneficiary 'for filing'. In this event, one must not omit to specify that the confidentiality and nonuse obligations still apply to this copy until it is returned or destroyed.

Finally, one must not forget that the confidentiality contract is actually the framework for negotiations. Consequently, the manner in which these negotiations are to progress and, above all, the outcome thereof must be foreseen in the contract. How much time is allowed for discussion? When and how will the beneficiary express their interest or lack of interest in the contemplated technology? What type of contract is contemplated if the parties confirm that they are interested in pursuing their relationship (license, assignment, partnership, etc.)?

As start-up companies, particularly when they are formed, often tend to sign a number of NDAs without keeping specific traces of them, this may pose important problems, particularly at the time of industrial property due diligence, as some of these contracts may possibly not be found and it is therefore necessary to remind them that, even if a technology transfer contract is subsequently signed, the NDAs must remain.

In particular, when the NDA does not give rise to a contract, the non-use clause may become very restricting and may block the development of the start-up. In the case of academics, one must be careful about the non-disclosure clause, as a scientist may not be able resist a publication in *Nature* or *Science*, even if his organization has signed a secrecy agreement.

In view of the exponential development of contracts in the life science sector and the size of the parties concerned, public bodies or 'big pharma', it is not impossible to be confronted with a conflicting requirement – a non-confidentiality agreement. In this case, the two parties recognize that the information exchanged will be able to be used by the parties within the limit of the right that they may be given by the filing of a prior patent application. This type of clause allows preliminary exchanges without subsequent obligations.

2.5 Preliminaries to Negotiations for a Technical Transfer Agreement

Once the initial contacts have taken place, a first actual negotiating session follows a number of intermediate meetings, each of the parties having already become quite agitated about these often irrelevant preliminaries. This is why, when the parties arrive at the substantive negotiation in respect of the transfer contract, they are often under psychological conditions that may prove to be difficult.

It is always preferable to negotiate a contract with a very crafty lawyer who specializes in contracts, rather than with an intelligent amateur who has no concept of what technology transfer really is. It will therefore always be preferable to have a very bitter argument with a professional about the various clauses of a contract rather than imagining that a negotiation with a scientist who thinks they are a lawyer will possibly enable one to obtain better terms. This is a delusion that will waste a lot of time and very frequently the terms finally achieved will be so convoluted, so far outside the usual norms, that they will become very difficult to apply.

The negotiation is followed by simplified agreements known as a Memorandum of Understanding (MOU) or a Letter of Intent.

2.6 Memorandum of Understanding or Letter of Intent

The next stage is to negotiate and sign an MOU or Letter of Intent. It is a matter of putting down on paper the main principles on which the parties agree in order to facilitate the work of the persons drafting the contract itself. If this document is not available, it is fairly futile to draft a contract.

However, it should be borne in mind that small companies too frequently fail to understand that some Letters of Intent are binding, meaning that they involve clauses that must recur in identical form in the contracts. It is therefore necessary to convince them that negotiating MOUs or Letters of Intent under these circumstances may be as difficult as the contractual negotiation itself.

A Letter of Intent or MOU must include at least the following:

- The parties in question.
- The type of contract, license, assignment, 'R&D'.
- The domain.
- Exclusivity and right of sublicensing.
- Financial consideration, at least in the form of a price range.
- In the case of 'R&D', at least a research agenda and the corresponding financing.

If a point causes problems during the initial exchanges, one must not forget to mention it in the MOU. If the negotiation has to be abandoned, one might as well do so straight away.

The MOU may take the form of a table in which the two parties' proposals are set out. If they do not agree, each party can then assess its room for maneuver for the next meeting.

2.7 Material Transfer Agreements

Strictly speaking, these are not generally preparatory contracts. They are special contracts permitting the transfer of material in general (biological material: cells, cultures, microorganisms), essentially for research purposes first of all, but they may be a prelude to a Technology Transfer Agreement (TTA).

It must be considered that Material Transfer Agreements (MTAs) have been criticized for a long time in view of certain clauses that prohibit commercial use and the fact that, if patentable inventions are developed, the relevant patent applications would be filed in the name of the company supplying the material.

Under these circumstances, some start-up companies have involuntarily developed research which they themselves were unable to exploit when it was completed, but which was the subject of patent applications in the name, for example, of foreign universities!

When signing MTAs it is therefore necessary to ensure that they do not include this type of clause so that one does not risk damaging the viability of the company in the future or else to be aware of the limitations of this type of contract. For example, if the material is used only in order to carry out checks, there is no great risk; on the other hand, developing products on the basis of material covered by an MTA is a risk that must be properly evaluated.

Ideally, the party whose material is transferred to the other party will expressly provide that the MTA does not cover the transfer of ownership or a license to the rights in respect of this material. Consequently, this material cannot form the subject, in whole or in part, of a patent application by the party benefiting from the transfer, and it may not be used for industrial and/or commercial purposes.

It is also necessary to tackle the question of the guarantee relating to the material transferred. Where the material involved is experimental, and therefore likely to be developed and not yet set as regards all its properties, the party making the transfer will take care expressly to exclude any guarantee, both as regards the safety of this material and whether it is appropriate for the use intended by the party benefiting from the transfer, or else as regards the IP rights that third parties might claim in respect of this material.

A number of sites, particularly those belonging to the US National Institutes of Health, supply standard wording for MTAs that is now widely accepted by the biotechnology community.

It must also be appreciated that this type of contract may be required when preparing scientific publications, particularly when it is necessary for living material to be made available to the scientific community to enable the results described to be reproduced. This type of technology transfer must again be managed with many precautions against the risk of part of the company's know-how being made available to a scientific community, a great majority of whom will respect their obligations; however, in the case of a very small minority, there is a risk of the material being used under dubious circumstances. The party transferring the material is recommended to make a prior check of the content of the contemplated publications and to place its name on the published documents.

In this case also, it is necessary to locate these contracts which are often regarded by scientists merely as letters or 'order forms', whereas they are real contracts and must therefore be conserved and analyzed if applicable at the time of a due diligence.

2.8
Founder Contracts in Technology Transfers

The initial contracts that we regularly face at the European level in the sector of state-of-the-art technologies are, in general, contracts with public bodies. This is because a large number of small emerging companies are formed as spinoffs from university research or other state bodies (e.g. institutes). Although these contracts are not, strictly speaking, technology transfer contracts, they are contracts that may have a fundamental impact on future negotiations.

For this reason, when TTAs are made with companies of this type, it will sometimes be necessary to verify the exact status of the transferred technologies, in particular patents and know-how in respect of which the companies have the following rights:

- Ownership.
- Joint ownership (the most frequent situation when different public bodies or universities have been involved).
- Exclusive or non-exclusive licenses.
- Sublicenses.

Providing 'founder contracts' with relevant clauses may be a prelude to a TTA negotiation.

It is necessary to check, particularly in the case of ownerships or joint ownerships of patents or patent applications, that the company was indeed entitled to file in its own name or that the proprietors were indeed entitled to the patent and that none of the inventors is in an 'irregular' situation, that is to say having no legal connection with the joint proprietors and that, having apparently been abandoned as a proprietor, he is capable of appearing at any time to dispute the transfer (in particular by means of proceedings claiming ownership).

Depending on the country, the situations of inventors working for public bodies may vary greatly; they may even vary depending on the universities or the bodies. It is therefore always necessary to check not only that the situation of the inventors, as between themselves, has been settled, but also that this situation is in accordance with the legal or customary provisions of the bodies to which they belong. A specific guarantee clause on this point may be justified.

For example, the situation in France relating to doctorate students is still not completely settled [see the case of *Puech v. CNRS* (Centre National de la Recherche Scientifique – a public research organization)] and although it had for a very long time been thought that their rights had to be transferred *de facto* directly to the body or university where they prepared their thesis, this interpretation is currently being challenged and a recent decision of the Court of Cassation relating to a CNRS intern could have a substantial impact on a number of previous contracts.

Michel Puech, a doctor and ophthalmologist, had been an unpaid intern during the 1996–1997 academic year at the LIP laboratory (Laboratory of Parametric Imagery of the University of Pierre and Marie Curie), an entity subordinate to the CNRS. His training, carried out under the direction of CNRS researchers, concerned studying the improvements that could be made to an examination of the eye by means of high-frequency echography apparatus. In the course of experiments, it became apparent that the use of this apparatus was interesting not only for the purpose of studying the front part of the eye, which was already known, but also for the rear part, which was not known. From that time, Mr Puech changed the orientation of his Doctoral thesis, which he submitted to the Compiègne Technical Institute in September 1997. Mr Puech also filed a patent relating to the result of the experiments carried out on the observation of the rear segment of the eye.

Proceedings were then brought against Mr Puech, first by the CNRS researchers with whom he had worked, who claimed paternity of the invention, and second by the CNRS, claiming the ownership of the patent. The District Court dismissed the plaintiffs' claims and ordered the CNRS to pay Mr Puech the sum of EUR 155 000 by way of damages for abuse of process and improper pressure.

Before the Court of Appeal, the appellants invoked, in particular, the LIP service regulations that give CNRS the ownership of patents resulting from work carried out in that laboratory. The Court of Appeal decided that Mr Puech was indeed the only inventor. However, it took the view that the patent lawfully belonged to the CNRS under the service regulations governing the relationship between Mr Puech, as a user of the administrative public service provided by CNRS by means of its laboratory, and the public body that was specifically the CNRS.

The Court of Cassation quashed the judgment of the Paris Court of Appeal, stating that the industrial property right belonged to the inventor and that only an exception provided for by law could derogate from this principle. However, none of those exceptions applied in the present case, as Mr Puech was not an employee or a public official. Consequently, the Court of Cassation held that Mr Puech was not only the inventor, but also the only owner of the rights in respect of the invention, namely the patent.

It is therefore necessary to ensure, in contracts with persons who are not employees of the company, that an IP clause is provided which deals with the ownership of the results developed by the non-employee (intern, doctorate student or postdoctorate student).

Likewise, in the case of negotiations with start-up companies, it is essential to check the position of researchers with regard to the company. This is because some national provisions permit researchers who are civil servants to participate directly in the management of the company, while other countries do not permit this, and in some countries it is a criminal offence. Contrary to what one might think, most researchers are completely unaware of this aspect and do not hesitate in any way to form their own company in breach of their obligations as a civil servant. In addition, the status of a consultant may give rise to problems, including for Scientific Advisory Board members. Of course, any negotiation carried out under these circumstances may lead to considerable problems.

Where the company's right is a license in respect of the technology of the invention and a third party wishes to obtain a sublicense, it is necessary to check that this sublicense is authorized; however, above all, it is necessary to ensure the 'continuity' of the sublicense. This is because a number of events may affect not only the licensee and 'second-rank' licensor, but also the main license contract. If the licensee and 'second-rank' licensor were to disappear, or if it were merged or purchased by a third party, the sublicense contract would, in the absence of such a clause, become null and void, and the sublicensee would then lose the benefit of the sublicense contract. Likewise, if the main contract were to be terminated (e.g. because of cancellation by the licensor for non-payment of fees by the licensee) the sublicensee would no longer be entitled to benefit from the right to enjoy the licensed patents, by virtue of the principle that 'no person may transfer more rights than he possesses himself' (*nemo plus juris ad alium transfere potest quam ipse habet*).

It is therefore essential to make provision, by means of a specific detailed clause, for the outcome of the sublicenses in these two events. Provision is frequently made for the main licensor to 'take over' the sublicenses, either as they stand until the final dates thereof or on terms which have to be renegotiated. This is because biotechnology start-ups often tend to disappear, either because of failure or because of absorption or other financial transactions, or very often when the sublicensee finds themself in an uncomfortable situation. They no longer have a licensor and are uncertain whether the patent proprietor can take over the licensee's obligations (e.g. technical assistance, know-how) or even whether the patent proprietor is obliged to take over the rights and obligations on identical terms.

This clause must be carefully studied depending on the identities of the patent proprietor and the licensee.

2.9
The Technology Transfer Agreement Contract Itself

It is not the purpose of this chapter to review all the clauses that may occur in TTAs, but rather to concentrate on the clauses that apparently cause frequent problems in the life science sector, in view of the very particular nature of the parties concerned.

Simple technology transfer contracts can essentially be of two types:

- R&D contracts with an operating license.
- Operating licenses covering one of the patents and/or the know-how (possibly ancillary rights).

Contracts for the provision of services or assignment contracts do not pose specific problems other than the ownership of the rights in the case of an assignment and the obligation for the change of proprietor to be officially registered in the case of assignments.

A contract that is not directly derived from the TTA, but which may have an impact on the transfer, is the joint ownership provision in the case of a license relating to a patent that has a number of proprietors. The document will have to be disclosed before the TTA is signed in order to ensure that no clause prevents the granting of a license.

We will consider below the problems raised by the part of R&D or license contracts relating to utilization.

The R&D contract is generally made up of a research contract that has no special features in the life science sector and a utilization part that has all the characteristics of a license. The only point that has to be remembered in the R&D contract is that it is not desirable to postpone the utilization part until a subsequent negotiation, which can only create problems for the future. The minimum requirement is to provide a kind of MOU that defines the utilization conditions if the research is successful. It is a mistake to fail to provide for this phase, even if it is always a little frustrating to negotiate something which does not yet exist, but one must be optimistic!

2.10 Ownership of the Rights

In the life science sector, one must always be very attentive with regard to the ownership of information and patents that are transferred, particularly when they originate from public bodies. In general, this type of research is carried out as a result of interaction between a number of public or quasi-public laboratories, each of which has individual right, and inventors who belong to it or who are paid by third parties. Finally, attribution rules exist and these may vary depending on the bodies. A preliminary analysis of the rights is always preferable to an unpleasant surprise in the long run (see also Section 2.8).

The various parties involved very frequently refuse to perform this preliminary analysis on the grounds that everything has gone very well up to now. It is very well known that problems arise only when the project is developing favorably, but by then it is often much too late.

Additionally, in the case of negotiations with a group of joint proprietors, universities or institutes, one must not forget to obtain confirmation that the person one is dealing with is duly authorized to negotiate on behalf of all the parties involved.

2.11
Subject Matter of the Contract

Most of the time, the license will cover one or more patents and know-how, possibly 'living materials'. In terms of product, life sciences have a singular feature, which is that the product covered by these contracts is very often matter which reproduces itself or which can easily be reproduced. For example, one can imagine that assigning two animals may lead to different contracts depending on whether they are:

- Two male mice.
- A pair of male and female mice.

The structure of the contract therefore needs to be carefully analyzed every time in order to determine the precise terms, in particular the financial terms, of the contract as well as the limitations to be introduced to take account of selfreproduction. It must also be borne in mind that know-how may very often be of greater importance or at least the same importance as patents in terms of implementing the transferred technique. This may justify a division of the royalties between the patents and the know-how, which makes it possible to pay royalties even in countries where there is no patent protection.

Laboratory notebooks, which are both essential elements of the transfer and precious documents in defense of the right, particularly in the United States in the event of interference, must never be forgotten in the TTAs.

2.12
Domains

In emerging domains like biotechnology, the technique transferred is often a basic technique (e.g. the polymerase chain reaction patents). Under these circumstances, the domain that is exclusively granted will often be limited and the patent or patents will generally be on a co-exclusive basis.

In this case, it is essential to realize that partitions, as they are called, namely the domains as being either 'A' or 'not A', the positive definitions of the term 'domain A', 'domain B', often lead to overlaps and experience proves that it is very often the case that research is done precisely in these grey areas which are covered by different contracts. For example, contracts which, *a priori*, are distinct, one relating to cancer and the other to the central nervous system, will very probably lead to a development dealing with cancers of the central nervous system.

Likewise, a domain that is very often used – the treatment of acquired immune deficiency syndrome (AIDS) – involves very varied angles of attack on equally varied pathologies that are directly connected. It is frequently difficult to determine a therapeutic area relating to AIDS that does not involve the immunological system, cell signals, apoptosis, etc. Under these circumstances, the implementation of inventions on a coexclusive basis risks leading ultimately to conflicts.

In that respect also it is necessary, from the start, to raise the problem of possible overlaps so as to provide for a way to resolve problems and not allow problems to develop. Ultimately, apart from exceptional cases, R&D in more than two distinct domains on a co-exclusive basis will probably lead to problems in utilization, but also in case of infringement lawsuit.

2.13 Territory

In the biology field, it is quite usual for the territories put in place to be the whole world. It is equally common for the company, often small in size, to possess patents only in the main countries capable of receiving this type of development, namely Europe, the United States, Japan and Canada.

Under these circumstances, it may be questionable to pay or to arrange to pay royalties in countries where no form of protection exists. For this reason, it is always necessary to provide for an express apportionment of the royalties, corresponding, on the one hand, to the licenses in respect of patents and, on the other, to the know-how part, in the knowledge that only the latter part will be paid when the product or process is utilized from a country where no patent exists to a country where likewise no patent exists.

2.14 Know-How

When one just reads contracts, know-how very often appears to be the poor relation of the patent. Although it is present in the great majority of technology transfer contracts, it is generally poorly defined and an even more undesirable fate is reserved for the terms under which it is to be delivered, which generally do not exist.

Know-how generally means the totality of the processes, recipes, methods and, more generally, knowledge that is generally not patentable (but this is not exclusively the case – the secret may also be retained as a mode of protection), which are secret and form part of the product manufacturing process, and which confer added value on that process and/or on that product.

A more legal definition, which has the merit of being both concise and clear, is given by European Commission Regulation No. 772/2004, which is referred to hereinafter. Know-how means 'a package of non-patented practical information, resulting from experience and testing, which is secret, substantial and identified'.

Very special care must therefore be given to the drafting of clauses relating to know-how, which, as previously mentioned, is the consideration for royalties relating to the utilization of the technology in most countries in the world. First, the definition of the know-how, which must refer to a written and confidential document containing the characteristics of the know-how in the most precise and exhaustive manner possible. Then there is the communication or delivery of this know-how, which will take place in the form of the delivery of documents or even by means of the secondment of staff in possession of this know-how for a more or less long period. The licensor will ensure that it receives in writing an acknowledgement by the licensee that the know-how has been duly and validly communicated to him and that it is now able to utilize it fully, as well as the patents. Failing this, the licensee, who is generally subject to a minimum utilization obligation, may consider that it has not been sufficiently put in a position to utilize the technology correctly because of insufficient delivery of the know-how.

2.15
Financial Considerations

The next problem, of course, is the problem of financial considerations, which is hardly an original subject as it is at the center of all contracts. Nevertheless, in the biotechnology context, there are a number of specific transactions.

The concept of milestones – the payment of inclusive sums prior to the utilization phases – was generally rejected by companies, including the pharmaceutical companies, in the 1980s. Things changed with the appearance of start-ups in biotechnology. Indeed, how could one imagine that such fragile companies could wait 10 years (the average period for the development of a drug) before receiving any money. That would simply condemn them to disappear. For this reason, milestones have become essential and are now widely accepted in this type of negotiation, but also, generally, in other contracts.

As regards the payment of milestones, one very often has the impression of being in a game of poker where practically none of the players really knows the rules of the game. A figure is put forward and it is then discussed. Very often, the start-up company ignores the contract itself and takes more account of the sums that it wishes to have in order to continue its development.

First of all, one must not confuse milestones with the possible development costs, which may be calculated relatively objectively. The milestone is regarded as being a reward for carrying out a development stage. It is therefore obviously left to the parties to assess, but it must also to some extent be compatible with the development costs. This is because it is clear that a company that has agreed to invest substantial sums in order to achieve a stage in the development must be prepared to pay relatively high sums when the positive results are obtained. From another point of view, that company may consider that it has already made great efforts in the matter.

It is also important to remember, particularly in this type of payment, that a number of products may be developed concurrently with overlapping clinical phases and that accumulations of milestones may arise that must be taken into account when drafting the contract.

It is necessary to be careful when contract values are mentioned, because a deal worth EUR 30 million, generally corresponding to the consolidation of all the milestones, very often terminates with a milestone of EUR 25 million at US Food and Drug Administration authorization, in other words in 10 years, which greatly reduces the size of the deal, particularly for investors who of course intend to realize their shares after 4–5 years.

In addition to the possible down-payments on signature, the purpose of which is generally more or less to repay the investments already made, the milestones are generally scheduled in accordance with the development phases (i.e. preclinical, phase I, phase II and phase III) and in certain cases other phases are retained.

It is therefore necessary for the end of these various phases to be clear. Merely stating 'for example when the licensor considers the results to be satisfactory' may lead to a product being developed over several decades. It is therefore necessary for the results to be calculated as far as possible. For example, provision can be made that the sums are due on 'entering the phase' or when the first patient is included in the phase in question in a given country.

2.16
Financial Clauses

Financial clauses are generally difficult to negotiate because the various parties involved often have fundamentally different ideas about the contractual area:

- Public research bodies.
- Start-ups.
- Major pharmaceutical companies.

Although percentages in terms of royalties generally do not cause many problems, parties very often forget to define the basis of these royalties clearly. In the biology field, one is dealing with a special phenomenon – the marketed matter is capable of reproducing itself. For example, the situation is different if, in the case of animal specimens, one sells two male rates or a male and a female rat. In that case it is often difficult to assess the corresponding basis clearly.

Likewise, companies carry out research that for some of them may be of a quasi-commercial nature and it may be a question of services rendered. In that situation also, assessing the basis will raise problems.

It is not possible in the context of this chapter to study the various relevant cases; it must simply be considered that the definition of the basis must often be made in conjunction with the definition of the product or the process used, carefully analyzed to take the various relevant cases into account. In some cases, recourse may be had to systems of tables which alone make it possible to take account of the various elements covered by the contract.

The royalty rate varies very greatly depending on the state of the project; for a therapeutic product it will be 4–5% before or during the preclinical trials and up to 10–15% or more in phase III.

In general, provision is made for royalties to be reduced depending on the volumes or turnover achieved. On the other hand, however, provision may be made in some cases for an increase in the royalties. The reason is that, although it is onerous to prepare a pair of transgenic mice, the cost of producing thousands of mice is very limited and therefore the margins are greater and so the royalties may be higher.

Strictly speaking, there are no other specific features of the sector, save that all strategies are still possible.

In actual fact, biotechnology start-ups are valued at least as much in terms of money as in terms of contracts signed with recognized companies or collaborations with research bodies. It is necessary to consider everything which may ensure signatures on technology transfer contracts or R&D contracts. This is also the reason why the first contracts signed in respect of the Cohen and Boyer patents made provision, which was quite original at the time, for down-payments to be paid when the contract was signed but with an advance payment in respect of royalties which is double the amount of this down-payment. The reasoning was that if the patent was not utilized by the licensee, it lost the amount of this down-payment but if, on the contrary, it utilized the patent, it was Genentech that would lose that sum.

Since then, other original strategies have been set up, but it is still possible to be creative in this respect.

2.17 Improvements

Improvements constitute one of the clauses that is the most difficult to manage in the biology field. This is because contracts are in principle frequently signed before any development whatsoever occurs. Accordingly, in order to set up a project that can really be used (e.g. start of the clinical phases), it will be necessary for quite a large amount of research to be carried out on both sides. Under these circumstances, it must be feared (or hoped) that important developments will be made that will be essential for utilization purposes. The resolution of this problem therefore constitutes an essential element of the contract.

It is fairly common to incorporate improvements in contracts, with or without amending the terms. It should, however, be recalled that the actual definition of improvement may be open to discussion. There are two overall approaches:

- A practical approach, consisting of regarding as an improvement anything that is supposed to improve the implementation of the product or the process
- A legal approach, consisting of regarding as an improvement anything that cannot be utilized without infringing licensed patents.

The first approach, which appears the more satisfactory, is of course very extensive and may cover a large part of everything that will be done by the various parties to the contract. This, of course, will be a problem in the future.

The second, legal approach has the drawback of not being stable over time, at least at the start, as it is necessary to wait practically until the patents are granted in order to know exactly what will or will not depend on the claims that are granted, bearing in mind additionally that these claims may vary from country to country.

Some decisions in Europe have also been innovatory in this area, bearing in mind that economic innovations that are important for the utilization of the contract, for example a new process, could be regarded as an improvement even if, *a priori*, there was no direct dependence.

The provisions of the exemption regulation, Commission Regulation No. 772/2004 of 27 April 2004, which came into force on 1 May 2004, apply in particular in this very crucial area of improvements. It should be recalled that this regulation applies when the relevant contract is capable of 'having a substantial effect on Community trade', according to the combined terms of the Treaty of Rome and of the European Commission. This criterion of substantial allocation is deemed to be met when, in the relevant Community market for goods or services, the total market share of the parties is 5% or more and the average annual turnover achieved in the Community by the undertakings in question reaches or exceeds EUR 40 million. In some very restricted markets, these thresholds may easily be met.

It should also be recalled that these criteria apply independently of the law applicable to the contract or independently of the law under which the undertakings that are parties to the contract were formed.

As regards improvements, clauses obliging the licensee to do the following are automatically excluded under the Regulation from the benefit of the exemption and they are therefore presumed to be anticompetitive and eligible for annulment in that respect:

- Clauses obliging the licensee to grant an exclusive license in respect of severable improvements or in respect of the new applications originating therefrom.
- Clauses obliging it to assign these improvements or new applications.
- Clauses obliging it not to contest the validity of the patents granted, unless the clause provides for the penalty for contesting such validity to be limited to the mere cancellation of the license contract.

It should be clarified that, in the Regulation, 'severable improvement' means any improvement that 'cannot be exploited without infringing the licensed technology', which, in our view, means inventions that cannot be commercially and/or industrially exploited without infringing a dominant patent, which comes back to the legal definition referred to above.

It must be appreciated that the large research bodies in Europe are very reluctant to grant a very extensive improvement clause because they do not know (!) with certainty about all the research carried out in all of their laboratories. The clause is therefore often limited to those improvements carried out in a specific laboratory or department.

2.18
Rights of First Refusal

In this respect also, when a large company in the life science sector signs a R&D agreement with a start-up, it wants to have the benefit of a right of first refusal in respect of the start-up's future developments or sometimes even in respect of future contracts.

It must be appreciated that this type of clause considerably diminishes the value of the company and may prevent developments, but, above all, blocks future negotiations. After all, what company would agree to negotiate and then see its proposal submitted to one of its competitors? In general, the very existence of this type of clause blocks negotiation with third parties from the start. It must in principle be completely controlled, particularly in the terms of the proposal, which must be specified, the broad outlines of the project, the general financial clauses contemplated and, in any event, there is no possible return after any refusal.

2.19
Circulation of Contracts

In a large number of European countries, contracts are entered into on an *intuitu personae* basis, in other words having regard to the identity of the other contracting party. The result is a fairly general rule that the contract may not be assigned to a third party without the consent of the other contracting party.

In the biology field and, in particular, in the case of start-ups, such a harsh clause is unacceptable. This is because a start-up company must in principle raise funds, change control and possibly be sold, and in these circumstances, as the assets of this company consist essentially of its industrial property and particularly its contracts, it is clearly impossible to develop a company if all the contracts cannot be transferred unless an outside company or companies give their consent.

Most contracting parties now accept this and a clause is substituted in which the refusal of the other contracting party's consent must be justified on the basis that this change of control or this transfer may be detrimental to the proper execution of the contract, in particular because the new holder of the contract does not have the capacity to exploit it or else because it has a product which is in direct competition.

2.20
Antistacking Clause

If there is any sector in which the antistacking clause (or stacking of royalties) is justified, it is the biology field. This is because there is currently practically no project that can develop without more or less depending on a number of previous patents.

There are now a large number of patents covering the basic technologies (which is normal in a technique which, in itself, is an emerging technique). Very frequently, most of the projects involve the need to have access to various basic or associated technologies (marking method, DNA duplication method like the polymerase chain reaction).

In a number of projects, the number of essential patents may be relatively high, sometimes as many as 10. Obviously, if each of the parties claims royalties of 3–5% in this case, the project will never be able to be developed. It is therefore necessary to provide a clause in the main contract that will enable some of the payments due to be postponed in order to pay any royalties to third parties, but this postponement must be limited, generally to 50% of the royalties that would be due were it not for those payments. This is the compensation clause.

In most cases, however, the exploitation of a project leads to an obligation to pay royalties to a number of patent proprietors and therefore to 'stack up' the royalties – this phenomenon is called 'stacking'. Although it is possible to provide for a certain cushioning of this effect with the previous comment, the problem in itself cannot be resolved when the patent holders are not parties to the contract and when compensation cannot totally compete. For this reason, when one speaks of an antistacking clause, it should be known that such a clause has a limited scope and cannot therefore resolve all problems of utilization in the future.

2.21 Various Clauses

In the event of negotiation, in particular with US companies, it is always necessary to take account of the cost of litigation proceedings. In view of the exorbitant costs of court proceedings in the United States, guarantees with respect to disputes in this country (obligation to defend or prosecute infringement proceedings) cannot be accepted or even contemplated by a start-up company. The only acceptable clause is one whereby the start-up company, as licensor, agrees to assist a licensee in the event of infringement proceedings, but it is not possible at any time to guarantee that the start-up company will bear the costs, or even part of the costs, of litigation proceedings of any kind in the United States. Such a request for a guarantee may become a term that causes the breakdown in contractual negotiation.

It is also important to retain a minimum degree of control over the patent proceedings that very often constitute a major part of the subject matter of the transfer. As a general rule, a potential exploiter of this technology must be able to obtain information and possibly intervene in the examination proceedings, in order ultimately to avoid becoming a licensee of patents in which the claims have been modified so that they no longer cover the exploitation intended by them.

In general, as was mentioned in the case of NDAs, confidentiality clauses, particularly when the contract is signed with a university, cannot be completely similar to the usual confidentiality clauses. If one wishes to work with scientists of a very high level, it is necessary to allow them to produce scientific publications, failing which they are very likely to refuse to sign the contract, particularly when it involves access to improvements or providing know-how. This is a clause that must be carefully negotiated in order to protect the interests of the relevant parties.

Likewise, in the life sciences sector, the duration of the contracts must take account of certain extensions that are specifically provided (e.g. supplementary protection certificates in relation to products such as pharmaceutical compounds or compounds intended to be used in the plant treatment sector).

Finally, it is always necessary, when drafting clauses that may put an end to the contract, to take account of the fragility of the start-up structures, which must have the benefit of a certain flexibility in the application of certain clauses, without those clauses constituting means for canceling the contract.

2.22 Conclusions

The preceding remarks do not of course constitute an exhaustive study of contractual obligations in general, any more than they take account of certain limitations connected, in particular, with competition rules in Europe. It should, however, be recalled that most contracts signed with start-ups in the life science sector, particularly in the contexts mentioned above, often have a global impact on the European market that is quite limited.

The key problem most frequently encountered nowadays in the life science sector is that of freedom to operate and thus the possible problem of royalty stacking; this is particularly clear in the biochip sector, in which it sometimes happens that several dozen patents may prove to be troublesome (e.g. patents filed in respect of each of the genes used in the creation of the biochip). In an attempt to solve these problems, the use of a number of patent pools on specific subjects has been proposed, or else it has been possible to create clearing houses permitting access to certain technologies involving numerous patents.

In the course of this chapter, it has not been possible to study in detail the structure of these contractual networks, but it must be considered that they are developing and it will be found in that respect also, when one wishes to use or implement the rules that result from establishing these networks (e.g. in the electronics sector), that a large number of adaptations are necessary in order to be able to use them in the life science sector.

3
Technology Transfer at the National Institutes of Health

Mark L. Rohrbaugh and Brian R. Stanton

3.1
Introduction

The National Institutes of Health (NIH) as an agency within the Department of Health and Human Services leads the US Government's support for biomedical research and training. The NIH is composed of 27 Institutes and Centers with more than 18 000 employees, and a fiscal year (FY) 2006 budget of US$ 28.6 billion.[1] Its mission is science in pursuit of fundamental knowledge about the nature and behavior of living systems, and the application of that knowledge to extend healthy life and reduce the burdens of illness and disability. Just under 10% of the budget funds the research conducted at the NIH (the intramural program) and just over 80% of the budget funds researchers outside the NIH, mostly at universities and hospitals in the United States but worldwide (the extramural program). It is estimated that NIH provides nearly 60% of US biomedical funding to US universities.[2] As the largest funding institution for biomedical research, the policies developed by the NIH to guide the conduct and management of NIH-funded research have a leading role in steering the activities of the biomedical research community.

Researchers funded by NIH, in both the clinical and basic research sciences, produce important new research findings, research materials and databases, advances in clinical care, and inventive technologies. The process of disseminating these results for the further advancement of science and, as necessary, the commercialization of technologies to meet public health needs may be considered under the broad umbrella of technology transfer. In this sense, technology transfer is not at all a new phenomenon. However, the manner in which such technologies are transferred, the role of the patenting and licensing of inventions, and the degree of commercial collaboration with academic and Government laboratories in this process has changed enormously in the last 25 years.

1) www.nih.gov.
2) http://www.nsf.gov/statistics/infbrief/nsf08320/.

Technology Transfer in Biotechnology. A Global Perspective.
Edited by Prabuddha Ganguli, Rita Khanna, and Ben Prickril

ISBN: 978-3-527-31645-8

This chapter will review the laws, regulations and policies that apply to the transfer of technologies from NIH-funded research, particularly the dissemination of research results, unique materials, and inventions. The authors will share perspectives on technology transfer policies and procedures that emanate from the experience of the NIH in its own technology transfer efforts. In addition, the discussion will include policy issues that have garnered the most attention and debate in recent years in the context of global public health challenges.

3.2 Technology Transfer Legislation

The transfer of technology from universities and Government laboratories is by no means a new phenomenon. However, decades ago, such activities were far more common in the physical sciences and engineering, which had more direct applications to industrial needs.[3] To the extent it occurred in the biomedical sciences, it usually involved diffusion of technologies through public disclosure rather than an active engagement or direct collaboration with the private sector of the research institutions with the commercial sector. However, some technologies, such as the polio vaccine, warfarin and cisplatin, invented before the 1980s were effectively transferred to industry. Prior to 1980, some agencies entered into Institutional Patent Agreements (IPAs) with individual universities to allow them to hold title to and license their inventions. While IPAs encouraged technology transfer, they created a system of unequal treatment of funding recipients sometimes with different, even conflicting, terms between different agencies and the same university.

All of this changed with the passage of the Patent and Trademark Amendments of 1980 (the Bayh–Dole Act)[4] and the Stevenson–Wydler Technology Innovation Act of 1980,[5] which established the modern era of technology transfer for extramural recipients of Government funding and intramural Government laboratories, respectively.[6] The intent of Congress was to promote US global economic competitiveness by addressing the lack of commercial uptake of Government-funded technology. The statutes provide incentives to research institutions to transfer inventive technology to the private sector for commercial R&D. In particular, the Bayh–Dole Act established a uniform patent policy for recipients of Government funding in granting them the right to elect title to inventions made under Federal grants and contracts.[7] This statute also strengthened the US

3) Rosenberg, N. and Nelson, R.R. (1994) American universities and technical advance in industry. *Research Policy*, **23**, 323–348.

4) Public Law 96-517. Although this statute only applies to non-profit and small business recipients of Government funding, President Regan extended it to large businesses under Executive Order 12591.

5) Public Law 96-480.

6) However, Federal laboratories were not given the right to enter into certain cooperative agreements with companies (see Footnote 14), retain royalties within their agency and provide the inventors with a share of the royalties until the enactment of the Federal Technology Transfer Act of 1986. Public Law 99-502.

patent system by consolidating eleven different appellate courts with jurisdiction to hear patent cases into one court – the Court of Appeals for the Federal Circuit. The expectation was that ultimately the US consumer would benefit with new products, new jobs and a more robust economy.

In exchange for the right to manage their intellectual property (IP) rights and keeping any royalties they earn, the funding recipients must favor small US businesses in their licensing efforts,[8] grant the Government a right to use the intellectual property 'for and on behalf of the US Government' worldwide on a royalty-free, non-exclusive basis,[9] i.e. a Government use license, require that licensees who manufacture a product for the US market manufacture the product substantially in the United States[10] and share some of the royalties with the inventors.[11] The Government also has the right to initiate 'march-in' proceedings under certain circumstances such as when the owner or the licensee of the patent is not bringing or does not have adequate plans to bring the technology to commercial application.[12] In addition, non-profit institutions cannot assign Bayh–Dole inventions to third parties without permission of the funding agency, except for an assignment to an organization that manages inventions as one of its primary functions.[13]

Congress has amended these statutes over time without substantial alternations in their structure, but has granted additional authorities to Government laboratories to conduct collaborative research under Cooperative Research and Development Agreements (CRADAs). Under this mechanism, the collaborating party and the Government laboratory can exchange personnel and materials, the collaborator can provide funds to but not receive funds from the Government[14] and the collaborator is offered an exclusive option to license inventions made by Government investigators in performance of the CRADA.[15]

7) The statute defines a subject invention as one which was conceived or actually reduced to practice in performance of the funding agreement. Note that the statute using the term 'contract' to refer to any research funding agreement, including grants, cooperative agreements and Government contracts under the Code of Federal Acquisitions, but excludes from these provisions other types of funding such as training grants. See 35 USC §201.
8) 35 USC §202(c)(7)(D), where 'small business' is defined as not having more than 500 employees. Small businesses, constituting the bulk of the workforce, were seen as engines of economic development.
9) 35 USC §202(c)(4).
10) 25 USC §204, with provision for a waiver process by the agency that funded the invention.
11) 35 USC §202(c)(7)(B). Note that Bayh–Dole does not set any particular amount to be shared with the inventors, whereas Federal agencies must share the first US$ 2000 and at least 15% of royalty income thereafter under a particular license with a cap per year of US$ 150 000 per person in total. 15 USC §3710c(a). Under NIH policy, its inventors share the first US$ 2000, 25% of the amount received above US$ 2000 up to US$ 50 000 and then 25% of amounts received thereafter in a given year.
12) 35 USC §203.
13) 35 USC §202(c)(7)(A).
14) Note that this is one of only four ways most agencies can receive funds, the others being Congressional appropriations, royalties from licenses and gifts funds, which can be restricted by the donor to a particular purpose, but not solicited by the Government agency nor accepted with any *quid pro quo* to the donor.
15) 15 USC §3710a.

Federal agencies exercise similar licensing authorities for inventions made by their scientists except that Federal agencies must limit exclusive licensing of inventions to those where such an incentive is needed for the licensee to invest the necessary capital to bring it to market. In addition, the scope of exclusivity is to be narrowly tailored to provide no more than the incentive necessary for the licensee to bring the invention to practical application.[16] Before a Federal agency can grant an exclusive or partially exclusive license, except for CRADA subject inventions, the agency must give public notice of the intention to grant the license and consider comments that are submitted in response to the notice.[17] All licensees must submit a development and marketing plan for the invention.[18] NIH uses this plan in part to develop the due diligence and performance milestones under a license, particularly for exclusive commercial licenses.

3.3 Impact of Bayh–Dole and Stevenson–Wydler Acts

Universities, Government agencies, and the business community by and large consider the Bayh–Dole and Stevenson–Wydler Acts to have been a great success in meeting the stated goals to enhance the transfer of technology to the private sector for commercialization. In 2002, *The Economist* concluded that Bayh–Dole was 'perhaps the most inspired piece of legislation to be enacted in America over the past half-century'.[19] Prior to Bayh–Dole, 28 000 patents resulting from Government-funded research were issued with very few licensed for commercialization. In 1980, US universities received less than 250 patents, but in 2004 they received 3800. More than 3100 products have reached the market since 1998 that result at least in part from university-licensed technologies. Since 1980, US universities have spun out more than 4500 companies, with two-thirds of these operating in 2004.[20]

At the NIH, technology transfer activities have grown significantly in the last 15 years. Royalty income has risen from several million dollars annually to US$ 97 million in FY 2008. The number of licenses executed annually has risen from 160 in FY 1995 to 259 in FY 2008. The portfolio includes about 3500 issued and pending patents, and over 1300 active licenses. Since 1987, over 400 NIH licenced products have reached market. While most of these are research reagents, 25 are FDA approved products, 17 are veterinary vaccines and one is a veterinary drug. These licensees have reported US$45 billion in sales from these products, with US$6 billion in 2007.

16) 35 USC §209.
17) 35 USC §209(e).
18) 35 USC §209(f).
19) *The Economist*, 14 December 2002 (US edn). There have been those who disagree or point out some of what they perceive as flaws. See 'Bayhing for blood or Doling out cash?, *The Economist*, 21 December 2005. Some of these articles are not completely accurate or neglect to include key facts. See www.autm.net. To the extent some of the problems are manifest, they represent the actions of a few institutions and not the technology transfer community as a whole.
20) AUTM Annual Survey 2004. www.autm.net.

3.4 Growth of Technology Transfer in Government and Academic Laboratories

A number of factors led to the expansive growth of the biotechnology sector in the 1980s. The legislative history and committee hearings prior to the passage of the Bayh–Dole and Stevenson–Wydler Acts suggest that Congress was most concerned with enhancing the economic competitiveness of United States in industries where it saw the technological lead slipping to countries like Japan and West Germany, namely those relying upon the physical sciences and engineering.[21] However, at the same time, the biotechnology revolution was giving birth to an entirely new industry. This entrepreneurial sector arose out of academia as distinct from traditional pharmaceutical companies, which produced small-molecule drugs and biologics processed from natural sources, including vaccines and proteins such as insulin and clotting factor. Ironically, prior to the passage of the Bayh–Dole Act, Drs Cohen and Boyer invented their recombinant DNA technology with funding from the NIH. The patent issued on 2 December 1980, shortly after the passage of Bayh–Dole.[22] Also supporting the development of the biotechnology industry was a decision of the US Supreme Court in 1980 that a genetically engineered bacterium was patentable subject matter.[23]

With the arrival of gene-splicing technology, researchers in the biomedical sciences found the more immediate results of their bench-top experiments of far greater commercial interest than ever before. Rather than being limited to their traditional role of laying the foundation for industrial drug design by elucidating the mechanisms of a biological function, biologist were now able to create genetically engineered microorganisms that could, e.g., produce commercially valuable proteins. Bayh–Dole and Stevenson–Wydler enhanced the importance of academic and Government research by providing institutions with new incentives and clear mechanisms to hold title to inventions, obtain patent protection, and the ability to use tools such as royalty-bearing licenses to exploit the commercial potential of new technologies (in this case, for public health benefit). It took several years before many public research organizations (PROs) would establish distinct technology transfer functions to capture technologies arising out of Government-funded research. The NIH itself initially managed patenting of inventions through the Office of General Council, moving this function over to the newly created Office of Technology Transfer in 1989.[24]

21) 1980 *US Code Congressional and Administrative News* (94 Stat. 2311), 4893; 1980 *US Code Congressional and Administrative News* (94 Stat. 3015), 6460.

22) US Patent 4237224.

23) *Diamond v. Chakrabarty*, 447 US §303 (1980).

24) The House Committee on Energy and Commerce, concerned with 'how to blend accelerated transfer with informed transfer', requested the Office of Technical Assessment to study technology transfer and assessment activities at the NIH. The report published in March 1982 focuses on the broader scope of technology transfer, primarily clinical trials and training to 'transfer research findings to the health care delivery system'. Only cursory mention is made to patents and licensing to industry in the comment that 'NIH is quite active in this regard, with approximately 370 patents licensed to industry'. OTA (1982) *Technology Transfer at the National Institutes of Health, A Technology Memorandum*, Congress of the United States, Office of Technology Assessment, Washington, DC, March, p. 52.

By the late 1980s, Bayh–Dole was hailed as a success with Government agencies and many research intensive universities having established offices dedicated to these technology transfer functions. However, it was not until the 1990s that many PROs began to see biotechnology technologies reaching the market yielding the first significant royalty streams.[25] Those who were not in the ballgame now wanted to play.

Organizations such as the Association of University Technology Managers (AUTM) grew significantly in membership and established models, training and facilitated the sharing of successful practices between members.

Long before the Bayh–Dole Act, scientists have had pressures, and sometimes acted upon them, to keep research results and important reagents from getting into the hands of their 'competitors'. By the 1990s, some of the first restrictions on the free flow of results of biomedical research appeared in the management of patent rights in a manner that had the effect hindering the progress of research, particularly with the use of research tools such as animal models, cell lines and antibodies. In 1995, AUTM and the NIH developed the Universal Biological Materials Transfer Agreement to facilitate sharing of materials between non-profit institutions.[26] The NIH developed internal policies favoring the licensing of research materials on a non-exclusive basis without obtaining patent protection.[27] After soliciting public and stakeholder input on hindrances to the exchanges of research materials, the NIH developed *Guidelines and Principles for the Sharing of Biomedical Research Resources*, known as the 'Research Tools Guidelines'.[28]

The Research Tools Guidelines require recipients of NIH funds to distribute materials that constitute research tools to researchers in all sectors – academic, governmental and for-profit. The terms of transfer agreement should not reach-through to capture rights in new materials made using the research tool, without charging for more than reimbursement for costs to researchers at PROs. In all of these policies, the focus is on using the patent system, and licensing in a manner that sustains and facilitates research while providing the appropriate incentives, including exclusive licensing as necessary, to the commercial sector for product development.

One of the specific challenges that arose at that time involved the distribution of Cre–lox mice, transgenic mice utilizing technology licensed to DuPont where the *cre* and *lox* DNA elements from bacteria are utilized in mice to facilitate re-

25) For example, the first FDA-approved product that included NIH patented and licensed technology was Fludara sold by Berlex after regulatory approval in April 1991. Between 1991 and 1995, the FDA approved six products that utilized technology licensed from the NIH. http://www.ott.nih.gov/about_nih/fda_approved_products.html.

26) www.autm.net.

27) See NIH Principles and Guidelines for Sharing Biomedical Research Resources. December 1999 http://www.ott.nih.gov/policy/research_tool.html, and Ferguson, S.M. (2001) Licensing and distribution of research tools: National Institutes of Health perspective. *Journal of Clinical Pharmacology*, **41**, 1075–125 and Rohrbaugh, M.L. (2005) Distribution of data and unique material resources made with NIH funding. *Journal of Commercial Biotechnology*, **11**, 249–62.

28) http://www.ott.nih.gov/policy/research_tool.html.

combination of foreign DNA elements into the genome.[29] The NIH and DuPont entered into a Memorandum of Understanding (MOU) in 1998 to facilitate the distribution of mice for research purposes among non-profit researchers on a non-exclusive, royalty-free basis.[30] The MOU governed the transfer of Cre–lox mice to and from the intramural research program and served as a basis for the exchange of mice among non-profit research institutions because DuPont agreed to enter into agreements with these institutions 'in accordance with the terms' of the NIH/DuPont MOU. NIH entered into similar agreements with DuPont for 'oncomice'[31] and with the providers of human embryonic stem cells that were approved for use with Government funding.[32]

With the success and maturation of technology transfer operations, the public and Congress turned the question of the appropriate return to the taxpayers for their investment in NIH-funded research. The undercurrent of concern by the American public related to the cost and means for reimbursement for pharmaceuticals, primarily drugs. In 2001, the NIH responded with *A Plan to Ensure Taxpayers' Interest are Protected.*[33] The report notes that the greatest return to the public from NIH research is in extended life expectancy and reduction of disability such that, according to the US Congressional Joint Economic Committee, 'if only 10% of this increase in value is the result of NIH-funded research, it indicates a payoff of about 15 times the taxpayers' annual NIH investment'.[34] The report looked more closely at the 47 drugs with sales of more than US$ 500 million in 1999. Of these only four, Taxol, Epogen, Procrit and Neupogen, utilize technologies invented with NIH funding.[35] An additional study done by the Government Accountability Office confirmed that few widely-prescribed drugs on the market utilize patented technology made with Government funding. The study found that of the top 100 brand name drugs, on a dollar value basis, procured by the Veterans Administration or dispensed by the Department of Defense in 2001, only six and four drugs, respectively, utilized Government-funded inventions.[36]

These studies confirm that the primary role of NIH-funded research is to provide basic scientific knowledge and unique reagents to the greater research community. Companies often develop drugs and therapeutics based on this

29) US patent 4959317.

30) http://www.ott.nih.gov/policy/policies_and_guidelines.html.

31) Mice transgenic for an oncogene for use in cancer research, covered by DuPont patents US 4736866, US 5087571 and US 5925803. See MOU at www.ott.nih.gov/policy/policies_and_guidelines.html.

32) http://www.nih.gov.

33) http://www.ott.nih.gov/policy/policy_protect_text.html.

34) The Joint Economic Committee, US Senate, May 2000. The benefits of medical research and the role of the NIH, quoted in *A Plan to Ensure Taxpayers' Interests are Protected.* jec.senate.gov.

35) Epogen and Procrit are based on different uses of the same patented technology developed at Columbia University. Taxol was manufactured by Bristol-Myers-Squibb (BMS) utilizing a method of semisynthetic synthesis invented at Florida State University and is administered by a method invented at the NIH under a CRADA with BMS.

36) US Government Accountability Office (2003) *Technology Transfer: Agencies' Rights to Federally Sponsored Biomedical Inventions (GAO-03-536)*, US Government Accountability Office, Washington, DC, July. http://www.gao.gov/htext/d03536.html.

knowledge of biological systems. Even when a Government-funded technology is licensed for use in a commercial product, the licensee company most often receives an early-stage technology, and takes on the high risk and massive development costs to bring it to market. The technology licensed from a PRO is usually only one of several patented technologies that are used to manufacture or comprise part of the final product. Thus, the relative contributions of the PRO and the company must be taken into account in any discussion the contribution of publicly funded research to a marketed product.

Several times NIH has formally considered the issue of the role of NIH in the ensuring that drugs are 'reasonably' priced when those drugs arise in any way from NIH-funded research. As a reaction to Congressional concern about returns to taxpayers, the NIH adopted a policy in 1989 that there should be a 'reasonable relationship between the pricing of a licensed product, the public investment in that product, and the health and safety needs of the public'.[37] This 'reasonable pricing clause' was included in CRADAs and applied to exclusive licenses for NIH CRADA inventions. Industry reacted negatively to this clause and many companies withdrew from interactions with NIH. The NIH convened panels involving academic and Government scientists and administrators, patient advocacy groups, and industry to review the policy. The panels' recommended that the policy be rescinded because it created a barrier to relations with industry that did not serve the best interests of technology development. They viewed the benefits of rapid development of technologies for public health as so significant that they overrode monetary return considerations.[38]

In 2004, the NIH considered two requests to use its march-in authority based on what was viewed as excessively high prices for the drugs in the United States compared to their prices in Europe and Canada. One request related to Xalatan (latanoprost) manufactured by Pfizer for the treatment of glaucoma and based on technology invented at Columbia University with NIH funding. The other related to Norvir (ritonivir) manufactured by Abbott based on technology it invented with direct NIH funding. Two separate conditions that could warrant march-in were considered: (i) the patent assignee or licensee 'has not taken or is not expected to take within a reasonable time, effective steps, to achieve practical application of the subject invention' or (ii) 'action is necessary to alleviate health or safety needs which are not reasonably satisfied' by the patent assignee or licensee.[39] The march-in authority allows an agency such as the NIH to conduct an administrative proceeding similar to a trial to determine whether one of the statutory criteria for march-in is met. If the agency makes such a determination, then it can grant a license to the Government-funded patents to a new party or require

37) *A Plan to Ensure Taxpayers' Interest are Protected.* http://www.nih.gov/news/070101wyden.htm#references.

38) See Footnote 37.

39) www.ott.nih.gov/policy/policies_and_guidelines.html quoting 37 USC §203(a)(1), (2). The other prongs that would justify march in were not relevant here: (3) 'action is necessary to meet requirement for public use specified by Federal regulations....' and (4) action is necessary because of lack of compliance with the requirement in §204 for 'products embodying the subject invention or produced through the use of the subject invention will be manufactured substantially in the US'.

the owner/licensee to sublicense the technology for commercial development. With respect to Xalatan and Norvir, the NIH found that the statutory conditions that would support a proceeding for march-in were not met in that both products were on the market and widely prescribed by physicians such that the manufacturer had achieved practical application and met health and safety needs.[40]

Of particular note is the NIH interpretation of term 'practical application', which is defined in the statute as having been achieved when 'the invention is being utilized and that its benefits are ... available to the public on reasonable terms'.[41] The NIH concluded that 'available to the public on reasonable terms' was not a requirement for 'reasonable pricing'.[42] Moreover, the issue of drug pricing and the global implications was properly left to Congress to address, not the NIH, and that the 'extraordinary remedy' of march-in is not an appropriate means of controlling or regulating prices.

There are a number of challenges in considering how one would fully implement the march-in authority.[43] It is useful as a deterrent and action of last resort rather than a facile tool for forcing the owner or licensee of a technology to move toward commercialization. Moreover, licensing practice of PROs has matured in the last two decades. It is now common practice for a licensor to include specific diligence terms such that the license can be revoked if the licensee does not meet performance milestones in taking reasonable steps to commercialize the technology.[44] This is a far more effective tool to achieve the same end. In times of emergency when the public needs rapid access to a technology and a licensee is not able or willing to take necessary action, the Government has at its disposal the authority to use patented inventions, whether Government funded or not,[45] which gives a patent owner, as the sole remedy for infringement, the right to sue the Government in the DC Court of Claims for a reasonable royalty. The patent owner cannot obtain an injunction, receive compensation for lost profits or obtain punitive damages. The Government can also assert as a defense a license to the invention under Bayh–Dole if it was made with under a Government funding mechanism.[46] This remedy applies only to direct infringement by the Government or

40) http://www.ott.nih.gov/policy/march-in-xalatan.pdf and www.ott.nih.gov/march-in-norvir.pdf. Also, see Raubitschek, J. and Latker, N.J. (2005) Reasonable Pricing—a new twist for March-in rights under the Bayh-Dole Act. *Santa Clara Computer & High Technology Law Journal*, **150**, 149–167.

41) 35 USC §201(f).

42) A public meeting was held for the march-in request for Norvir. Their comments include those who supported this interpretation, including former Senator Birch Bayh, and those who spoke against this interpretation. See www.ott.nih.gov/policy/meeting/May25.htm.

43) See McGarey, B. and Levey, A. (1999) Berkley Technology Law Journal, 14, 1095–1116.

44) This would not be an option in the rare instance when an invention is made and commercialized by a company with direct funding from the Government, such as in the case of Norvir.

45) 28 USC §1498.

46) The Government's license under Bayh–Dole in which the patent owner grants the Government a royalty-free, worldwide license to use the patented technology 'for or on behalf of the Government' has been consistently interpreted by the Government as applying to the Government itself and its contractors, who are acting on behalf of the Government, but not to grantees, who merely receive funds under an assistance mechanism. However, there are no judicial opinions interpreting the scope of this license. See *Duke v. Madey*, 307 F.3d 1351 (Fed. Cir. 2002).

its contractors, with the authorization and consent of the Government,[47] rather than contributory infringement, for which the Government cannot be held liable.

As any program matures, it requires refinement of its policies to manage new challenges that come to bear upon the programmatic mission. By the mid 1990s, the NIH recognized that it needed a formal policy to guide the management of its patenting and licensing responsibilities for inventions arising out of the intramural research program. The policies are based on the general principle that the primary goal of technology transfer at NIH is ultimately the improvement of public health. Other factors such as obtaining a reasonable return in royalties under the license and the economic benefits to society from the creation of new technologies are important but always secondary to the goal of improved public health. Thus, the patent policy envisions the use patents as tools primarily when they are needed to protect the technology and provide an incentive for commercialization under licenses.

As a result, the NIH generally does not patent technologies that are only useful as a research tool, such as animal models, cell lines and drug screening protocols. When a technology has dual use as a research tool and a commercial product or service, the NIH will consider obtaining a patent for the technology. In licensing technologies, the NIH always reserves the right to grant research licenses to both for-profit and non-profit research. It can charge for costs associated with preparing and shipping materials but will not charge a license fee or assert its patents against non-profit researchers even if they are collaborating with a company in which the company has certain rights to the output of that research. The company requires a license from the NIH only if it is using patented technology in an internal research project or for a commercial product or service.

The NIH objects to the use of license structures that could unduly encumber future research findings and the use of new intellectual property. This includes the use of 'reach-through' terms to attach rights to the novel outcomes arising from the use of the licensed technology that is not covered by the licensor's patent claims. Such terms, for example, would include fees based on sales of a new drug discovered using a patented and licensed screening technology. Exclusive licenses are reserved for technologies where the commercial sector requires that incentive due to the high risk and large investment in bringing a technology to market. Even then, the license will be limited to a scope of the commercial interest of the company. In addition, the NIH always reserves the right to grant internal research use licenses even under exclusive commercialization licenses. These last two principles, or avoiding 'reach-through' terms and permitting further research, are important to providing an open research base free from significant encumbrances such a stacking royalties that would result from reach-through terms possibly hindering or making the commercial development of a technology financial undesirable.

Policies developed for both the NIH intramural and extramural recipients of funding, are based on these same principles of using the patent system to provide

47) 28 USC §1498(a).

constructive incentives for new products and services to improve public health and not for unnecessary encumbrances on the system. While general NIH policies may recommend against patenting certain types of technologies, such as animal models, which do not require greater incentives for commercialization, the policies are most importantly directed towards licensing activity. Patents *per se* do not create hindrances for research and commercial development unless they are enforced in a manner that has that effect. Of increasing concern as well is the use of contractual obligations for materials governed by patents so that undue restrictions that cannot be or are difficult to enforce under patent law are enforced under contractual agreements such as Material Transfer Agreements to transfer unique materials that fall within the scope of one or more patents.

3.5 NIH Efforts to Transfer Technology Globally

The focus of the NIH licensing and its policies is necessarily on promoting public health benefits for the United States. However, the public mission of NIH is global. In part, the United States has had humanitarian goals in mind in supporting research on diseases that burden primarily the developing world. In the last 20 years, US policy makers have affirmed that such research serves the US public indirectly in that infectious diseases that arise or are endemic in one part of the world can spread to the rest of the world. In addition, countries that are severely burdened with poor public health are less likely to become strong trading partners and stable democracies.

Similarly, the NIH has increasingly had global public health in mind in licensing technologies of importance to developing countries.[48] For technologies with a potential impact on public health needs worldwide, the NIH has required licensees to provide plans for brining the product to market in at least some developing countries either concurrent with or subsequent to market approval in Western countries. In addition, technologies have been licensed directly to institutions in developing and emerging-market countries that the capacity to manufacture drugs or vaccines. Technologies for dideoxyinosine, and vaccine technologies for rotavirus, dengue fever, meningococcus, typhoid fever and vericella.[49] Another effort involves the collection of technologies related to neglected diseases invented by non-profit institutions and offered as available for licensing. The NIH currently hosts a website that lists technologies by disease and vaccine or drug categories with web links to the institution that owns the technology and would negotiate the license.[50]

48) Salicrup, L.A. and Fedorková, L. (2006) Challenges and opportunities for enhancing biotechnology and technology transfer in developing countries. *Biotechnology Advances*, **24**, 69–79.

49) Salicrup, L.A. and Rohrbaugh, M.L. (2007) Partnerships for Innovation and Public Health: NIH International Technology Transfer Activities in *IP Management in Health and Agricultural Innovation*. http://www.iphandbook.org/handbook/ch17/p12/.

50) http://www.ott.nih.gov/licensing_royalties/NegDis_ovrvw.html.

In addition to transferring technologies arising from the intramural program, the NIH believes that research institutions in developing and emerging-market countries need to be equipped to manage the technology transfer of their own inventions. To this end, the NIH has established a program for short-term training of individuals from such institutions.[51] To date, participants have included those from institutions in China, South Africa, India, Brazil and Mexico.

3.6
International Technology Transfer by Publicly Funded Research Organizations

Many countries look to the United States as a source for polices and procedures that can be adapted to address concerns in their localities. For example, the Organization for Economic and Cooperative Development (OECD)[52] leads initiatives that focus on harmonizing understanding and practices for trade-related issues. One of their initiatives is their guidelines for *Best Practices for the Licensing of the Genetic Inventions*[53] (the 'Guidelines'). This document represents the views of the OECD's 30 member countries regarding the licensing of nucleic acids, proteins, and methods of using these molecules in R&D. The Guidelines, largely emulating the NIH's Research Tools Guidelines,[54] globalize recognition of the importance of balancing the need for access to basic scientific information with the patent system's economic innovation incentive. The OECD Guidelines note that:

> ... over the last decade, as the number of such [gene-related] innovations has increased, their impact on health care has grown substantially. Recently, some governments, patient groups and healthcare providers have become concerned about how certain genetic inventions have, in certain circumstances, been licensed and exploited, particularly for diagnostic genetic services in the human health care field.

The Guidelines also note that:

> ... global issues remain regarding whether the intellectual property [IP] systems function effectively by encouraging the diffusion of information and technologies or [is] ... impeding access to genetic inventions ... [The Guidelines] conclude that the IP system ... functions largely as intended –

51) http://www.ott.nih.gov/about_nih/intl_tt.html.

52) www.oecd.org. 'The OECD groups 30 member countries sharing a commitment to democratic government and the market economy. With active relationships with some 70 other countries, NGOs [non-governmental organizations] and civil society, it has a global reach. Best known for its publications and its statistics, its work covers economic and social issues from macroeconomics, to trade, education, development and science and innovation.'

53) http://www.oecd.org/document/26/0,2340,en_2649_201185_34317658_1_1_1_1,00.html.

54) http://www.ott.nih.gov/policy/rt_guide.html.

> stimulating innovation and the disclosure of information, and that there is no evidence to suggest a systemic breakdown in the licensing of such inventions. Nevertheless, some specific concerns were identified, and in particular with respect to access to diagnostic genetic tests.

The Guidelines establish broad principles focusing on fundamental issues in the licensing of biotechnology including the importance of healthcare, research freedom, commercial development and avoiding anticompetitive practices. The guidance provided in the Guidelines took over 4 years to develop and is general in nature illustrating the time intensive nature of establishing even general global policy guidance. However, issues in technology transfer are highly fact specific and must account for the environment (legal, geographic and organizational) within which the technology is to be employed. Different actors presenting the public, private and non-profit sectors have distinct priorities, needs and constraints that must be considered when enabling technology transfer activities. These actors' conditions are further confounded by ethical, moral and social issues in the biotechnology industry because included among its many applications are pharmacology, diagnostics, and medical treatments. Each of these technologies is highly regulated and these regulations vary significantly across nations. Navigating the policy webs linking national, corporate and nonprofit communities is a difficult exercise, but linking these interests at one level or another are PROs.

International aspects of the interaction and collaboration among PROs remain of great interest. The success of Bayh–Dole within the United States is based on a variety of predicate assumptions including the particularities of the US patent system, more liberal market regulations in the United States, and the means by which the United States has implemented its obligations under international treaties including the World Trade Organization's (WTO) Trade-Related Aspects of Intellectual Property (TRIPS) Agreement and other treaties.

The WTO[55] is the successor to the forum associated with the General Agreement on Tariffs and Trade (GATT) that was established in 1947.[56] At the same time the WTO was formed, the TRIPS Agreement[57] was also negotiated and ratified. The TRIPS Agreement, ratified in 1994, was crafted in the shadow of the successful Bayh–Dole system and includes provisions that encourage a technology transfer environment similar to that of the United States. It is important to note, however, that the Bayh–Dole system, which arose as part of an evolutionary process, attempts to strike a coherent balance between 'pure' academic research

55) See the gateway to the World Trade Organization (WTO) that can be found at http://www.wto.org/english/thewto_e/whatis_e/whatis_e.htm.

56) GATT was first signed in 1947. The agreement was designed to provide an international forum that encouraged free trade between member states by regulating and reducing tariffs on traded goods and by providing a common mechanism for resolving trade disputes. GATT membership now includes more than 110 countries.

57) See the gateway to the TRIPS material on the WTO website at http://www.wto.org/english/tratop_e/trips_e/trips_e.htm.

that focuses upon 'philosophical speculation' and the practical adaptation of that research that leads to tangible public benefit.

The success or failure of any regulatory or legislative can be measured in many ways, but given the plethora of products and services based upon PRO technology[58] and the worldwide fascination with adaptation of the US Bayh–Dole/ Federal Technology Transfer Act acts to other national intellectual property legal landscapes, it is clear that these acts provide validated models for translating PRO research to the public. For example, one study indicates that, at least in regard to pharmaceutical development among US institutions, there is strong reciprocal relationship between the public and private sectors. This study examined the:

> ... interaction between the public and private sectors in pharmaceutical research using qualitative data on the drug discovery process and quantitative data on the incidence of co-authorship between public and private institutions. [It found] ... evidence of significant reciprocal interaction[s and rejected] ... a simple 'linear' dichotomous model in which the public sector performs basic research and the private sector exploits it. Linkages to the public sector differ across firms, reflecting variation in internal incentives and policy choices, and the nature of these linkages correlates with their research performance.[59)]
>
> Many current policy proposals and initiatives display the classic signs of international emulation-selective borrowing from another nation's policies for implementation in an institutional context that differs significantly from that of the nation being emulated.[60)]

Regardless of the adaptive mechanism, the international Bayh–Dole-type 'initiatives are based on the belief that university patenting was an essential vehicle for effective transfer of technology from universities to industry and that Bayh–Dole was essential to the growth of university-industry interaction in science-based industries in the United States during and after the 1980s'.[61)]

In Europe, while the majority of basic research is conducted by PROs, the route through which the results of their innovative efforts are translated into practical application has changed. As a general rule, European research has 'evolved from

58) *AUTM Licensing Survey: FY 2006 Survey Summary*, p. 10. A survey of 189 US institutions indicated that 697 new products were introduced into the marketplace and 553 new startup companies launched as a result of their technology transfer efforts. Association of University Technology Managers. http://www.autm.net/AM/Template.cfm?Section=Licencing_Surveys_AUTM&TEMPLATE=/CM/ContentDisplay.cfm&CONTENTID=2292.

59) Cockburn, I. and Henderson, R. (1996) Public–private interaction in pharmaceutical research. *Proceedings of the National Academy of Sciences of the USA*, **93**, 12725–30 and see Footnote 60.

60) Mowery, D.C. and Sampat, B.N. (2005) The Bayh–Dole Act of 1980 and university–industry technology transfer: a model for other OECD governments? *Journal of Technology Transfer*, **30**, 115–27.

61) See Footnote 60.

an open source model in which PROs did not retain any IP rights, to a 'Licensing Model' in which the PROs started to retain, protect and commercialize inventions based on their discoveries, essentially through licensing the IP rights to industry or to start-up companies'.[62] In the last 10 years, the European licensing model has been expanded to include an innovation model consistent with that in the United States. Whereas, in the United States, the lines between PROs and private industry have blurred as PROs spin-off private sector companies. In addition, personnel and their associated know-how pollinate private sector companies and industrial innovators often move to, collaborate with or provide resources to PROs.

Consistent with US findings, the European commission has found that a 'best practice is to vest initial ownership of results and inventions funded by public funds to the PROs where the research was conducted'.[63] They also noted that while spin-off company generation is more prevalent in the United States than in the EC, this is changing slowly and is considered to be a 'best practice'.

Translating the success of the US innovation model to the non-US communities remains a challenge as evidenced by statistics relating to, for example, European adaptation of PRO research to commercial technologies. Given the volume of ongoing research in European PROs relative to that in the United States, one could expect a 'far greater number of technologies being developed in an industrial context'.[64] However, this expectation may be unrealistic. The translation of US PRO innovation to practical application has been facilitated by technology transfer efforts that coming 24 years after the advent of the Bayh–Dole Act. These laws have only recently been introduced into the European communities, and it will take time for technology transfer systems to adapt and evolve from these changes to legislative and regulatory environments. It is clear that no single implementation model will suffice for all nations and the iterative adaptations necessary for the development of successful PRO technology transfer will take time.

Governments worldwide have sought to increase the rate of transfer of academic research advances to industry and to facilitate the application of these research advances by domestic firms since the 1970s as part of broader efforts to improve national economic performance in an era of higher unemployment and slower growth in productivity and incomes. In the 'knowledge-based economy,' according to this view, national systems of higher education can be a strategic asset, if links with industry are strengthened and the transfer of technology

62) European Commission (2004) *Working Paper on Community Research: Management of Intellectual Property in Publicly-funded Research Organizations: Towards European Guidelines*, European Commission, Brussels, p. vii. http://ec.europa.eu/research/era/pdf/iprmanagementguidelines-report.pdf. Note that the NIH does not work activity to establish new companies around its intramural technologies (i.e. spin-out companies) because it believes that this would not be consistent with its role as a Governmental agency that funds research primarily through grants and contracts to outside entities on a scientifically competitive basis. The NIH, however, does work to license technologies to start-up companies.

63) See Footnote 52.

64) See Footnote 52.

enhanced and accelerated. Many if not most of these 'technology transfer' initiatives focus on the codification of property rights to individual inventions, rather than the broader matrix of industry–university relationships that span a broad range of activities and outputs.[65]

For example, 'several countries ... have recently enacted laws, regulations or policies assigning ownership or the first right to ownership to PROs', including Austria, Belgium, Denmark, France, Spain and Russia. In the United Kingdom, patent rights have been vested in the university since the patent act of 1977. In other countries patent ownership has relied upon the so-called 'professor's privilege' system in which invention assignment vests in the professor or other public funding recipient. The latter systems were only recently rescinded in Finland and Norway and only remain in Sweden and Italy among European Union Countries.[66] The criticality of the difference between these two modalities should not be underestimated. For example, one analyst notes that:

> ... until recently, German universities were not interested in dealing with intellectual property issues because, by law, professors retained ownership of their discoveries. As a result, universities saw little return from licensing patents to companies. This all changed in February 2002 when a new law came into force that shifted intellectual property ownership to the universities and ruled that academics are to receive 30% of the licensing revenues.[67]

Since the introduction of the changes to section 42 of the German Employed Inventor's Act, the Max Planck Institute reported licensing revenues in 2003 of DM 32 million and Bernhard Hertel, managing director of the Max Planck Society's (MPS) technology transfer division, says that, '... there is an increasing demand from young scientists who want to start their own companies, not only at MPS but elsewhere in Germany'. Germany also maintains a program called 'EXIST' that promotes 'networks between universities, capital providers, and service companies to facilitate university spinouts'.[68] In still other countries, such as Denmark, patent rights are split between the university and the faculty member.

Regarding yet another example, Goldfarb and Henrekson[69] opine that the:

> ... different incentive structures that academic researchers face in the United States and Sweden ... demonstrates that in Sweden academics face strong disincentives to take the time away from their academic pursuits to facilitate knowledge transfer to the commercial sector ... we believe

65) See Footnote 60.

66) http://www.eutechnologytransfer.eu/downloads.php.

67) Habeck, M. (2003) Humboldt University beefs up technology transfer. *Bioentrepreneur*, published online: www.nature.com/bioent/bioenews/112003/pf/bioent781_pf.html.

68) See Footnote 67.

69) Goldfarb, B. and Henrekson, M. (2003) Bottom-up versus top-down policies towards the commercialization of university intellectual property. *Research Policy*, **32**, 639–58.

> that it is unlikely that Sweden is harvesting the full commercial potential of its research output as successfully as the US.[70]

Other countries have still more varied intellectual property ownership schemes. For instance, while Italy has shifted ownership from universities to individual researchers, in Japanese universities ownership of IP rights resulting from publicly funded research is determined by a committee. In the UK and Canadian university systems, no single national policy governs IP rights ownership, although this is moving towards a system similar to that found in the United States.

Regardless of the mechanism by which ownership of PRO intellectual property is managed, there is a worldwide movement to vest interests in publicly funded research with the institution or person that has received that funding. The goal is to facilitate the university/industry collaboration that, for example, '... senior Japanese Government officials have declared ... [is] essential for Japan's economic revival'.[71]

In Europe, one report notes that:[72]

> ... the combination of weak intellectual property laws and expensive patent prosecution can be fatal to a country's intellectual property regime, as is the case in Spain. The EU [European Union] condenses all these problems into the following list of concerns. Poor EU performance could be explained by the culture of many EU research institutions. Problems cited included:
>
> - a continued over-reliance on a 'linear' approach to innovation, which assumed that investment in the supply side would automatically result in marketable innovations downstream;
> - measuring academic success on the basis of research papers or academic citations, with intellectual property creation, for example, often not given parity of esteem as a research publication;
> - peer review (and lack of external examination), which may tend to prevent academic networks opening up to external scrutiny; and
> - academics being given insufficient time, or promotion incentives to engage in commercial activities.

70) See Footnote 69.

71) Rutt, J.S. and Maebius, S.B. (2004) Technology transfer under Japan's Bayh–Dole: boom or bust nanotechnology opportunities? *Nanotechnology Law and Business*, **1**(3), article 8. pubs.nanolabweb.com.

72) Siepmann, T.J. (2004) The global exportation of the US Bayh–Dole Act. *University Of Dayton Law Review*, **30**, 209–43. http://law.udayton.edu/lawreview/documents/30-2/The US Bayh-Dole Act.pdf.

> The EU is vocal and specific in calling for reform of the research systems within its member nations and cites a litany of problems from 'poor knowledge transfer mechanisms from the science base to industry,' to 'significant barriers' within the academic culture itself that prevent commercialization. The EU also cites an overall lack of clarity among many member nations as to who actually owns intellectual property stemming from government-funded research.

Whether an invention is assigned to the innovator (person) or the institution (e.g. grantee), the process of obtaining patent rights and developing the partnership relationships through licensing or assigning rights that permit their translation into products and services is complex both legally and technically. The US experience has grown up over almost three decades and has involved exercises fraught with mistakes. Business acumen, patent and licensing experience are all needed for a successful application of PRO innovation for practical public benefit.

Actualization of technology from PROs to the public can, at least in part, be measured by formation of spin-out companies. A 'spin-out company' generally refers to an independent corporate entity that is created to exploit intellectual property. These companies provide means to gather funding, further educational and research efforts, and transfer knowledge between the public and private sectors. It also provides a means to reap financial rewards that motivate academics to pursue practical applications of basic research activities. However, the latter carries with it the danger that the lure of financial gain may shift the balance from the basic research enterprise to developmental activities carrying greater profit potential.

In the United States, a greater amount of public funds are used per spinout that in, for example, Canada and the United Kingdom. For example, 2001 data from AUTM and UNICO–NUBS[73] indicated that the United States spends approximately US$ 171 million for each spinout formed in contrast to only US$ 48 million in Canada and US$ 17 million in the United Kingdom.[74] The survey also:

> ... shows that during 2001 universities created 175 new spinout companies, accounting for 31% of all 554 spinouts formed in the [preceding] ... five years. However, much of the spinout activity is concentrated in relatively few universities. About a quarter of universities (26.7%) created more than 10 spinouts each but a quarter (25.3%) did not create any spinouts in this period.

73) University/Company Association (http://www.unico.org.uk) and Nottingham University Business School (http://www.nottingham.ac.uk/enterprise/unieihome_archive.htm).

74) See, e.g. 'Spinouts pick up speed'. http://www.hero.ac.uk/uk/business/archives/2002/spinouts_pick_up_speed2872.cfm.

Regardless of the system employed or the mechanism by which technology is developed, the ownership provided by Bayh–Dole type rights does not directly translate into IP rights and technological innovation. It is still necessary to have the requisite skill, policies and knowledge to obtain useful patent protection, and then the ability to utilize those IP rights to facilitate development of products and services. There are many factors that, in the United States, act as catalysts for translation of research results into product and services that directly benefit the public.

One cornerstone of the US economy is entrepreneurship and a permissive environment for, among other, translating early stage science into practical application. Derek Leebaert, a professor at Georgetown University, notes that:[75]

> Small businesses contribute much more to the US economy and society as a whole than can be calculated just from the spending and profit that they generate. These businesses tend to be more economically innovative than larger companies, more able to respond to changing consumer demand, and more receptive to creating opportunities for women and minorities, and activities in distressed areas. 'Building, running, and growing small business is a part of a virtuous cycle of creativity and increasing prosperity that can be applied by dedicated and thoughtful people anywhere,' the author says. 'There are no secrets, and frequently money is less important than a considered combination of imagination and effort.

Other factors that contribute to the ability of innovators in the United Nations to bring products and services to the consumer include access to a broad array of financial resources (including, for example, venture capitalists and Angel investors) and an open economic environment. In addition, the relatively unique aspects of the US patent system provide an environment that balances open information exchange against the exclusionary rights provided by the patent system.

In contrast to the rest of the world, the US patent system currently has a first-to-invent system rather than a first-to-file one. In the latter system, if there is a conflict between inventors claiming the same invention, the Government will grant a patent to the first party to file a patent application, presuming, of course, that all other conditions for patentability are met. In contrast, in a first-to-invent system the patent office will award the patent to the party that is able to demonstrate that they were the first ones to 'invent' that which is sought for patenting. Resolution of conflicts between parties seeking patents on the same invention is done through an expensive and complex process known as 'interference'. While discussion of interference practice is beyond the scope of this chapter, the reader should note that the complexity of determining who invented something first

75) Leebaert, D. (2006) How small businesses contribute to US economic expansion. *eJournal USA: Economic Perspective*, **11**(1). http://usinfo.state.gov/journals/ites/0106/ijee/leebaert.htm.

has driven a global transition to first-to-file systems. Whether this is helping or hindering innovation is unclear.

In the academic community, information exchange is largely done through the established system of publication in peer-reviewed journals. In the United States, this is usually where an invention is first disclosed and provides a means for broad disclosure of scientifically validated research results. In the first-to-file system, inventors need to make their first submission to the patent office which will not publish that information until 18 months after filing. It is only after filing the invention that an inventor becomes free to publish their research findings. Thus, the pattern of information disclosure is different in the United States than in other countries.

Another factor that contributes to the preservation of academic freedom and open dissemination of knowledge while preserving the potential incentives provided by the patent system is the 'grace period' provided by US patent law. In the United States, an inventor may disclose their invention to the public up to 1 year before filing a patent application without jeopardizing their potential patent rights. Similar types of grace periods are present in some countries while others (including members of the European Commission) have an absolute novelty standard that requires that patent application filing be the first disclosure of an invention.

Different countries address the so-called 'grace period' in different ways.[76] The absolute novelty standard best serves innovators that do not rely upon open publication for information dissemination (e.g. large industrial actors) and capital investment. In contrast, PROs rely upon peer-reviewed publications for information sharing and dissemination and keeping research results. Secrecy is anathema to the public research enterprise.

The potential importance and impact of the grace period on the ability to bring inventions to market should not be underestimated.

> [The] official view of the French and German Government as regards the introduction of a grace period in the European patent law, contains 10 points ... [including that the] introduction of a grace period in Europe would favor innovations, in particular a more rapid transfer of results of research and development into commercial application [and that] (r)esearch and scientific institutions would benefit at

76) The spectrum of 'grace periods' among countries can be divided into three basic categories: relative, local and absolute novelty. For example, Brazil, the European Patent Office, France, Germany, Mexico, South Africa, Taiwan, the United Kingdom and Venezuela have an absolute novelty standard for patentability. Any disclosure of the claimed invention to the public anytime before the filing of the patent application is sufficient to preclude patenting. In contrast, in some countries including the United States, Australia, China, Canada and Japan, there is a relative novelty standard that permits the inventor to disclose there invention to the public up to 1 year before filing a patent application without negating their ability to obtain patent protection. The third situation, local novelty, provides inventions may not be disclosed within the country of patenting prior to filing of the patent application.

> most, since the grace period would ease the conflict between an early disclosure and filing of a patent application. A grace period would be equally beneficial to small and medium size enterprises, in particular as far as their cooperation and public experiments are concerned.[77]

In addition filing and disclosure requirements, there is also some debate as to how 'new' an invention needs to be before it should be able to be patented. This is a global debate regarding the merit of 'incremental' versus 'evolutionary' technological advances. Incremental innovation provides a continuum of technological adaptation of preceding inventions whereas evolutionary standards provide that in order for an invention to be patentable there must be some 'flash of genius' or other substantive difference between that which is sought for patenting and that which has come before. This is especially contentious in biotechnology and pharmaceuticals where minor advances that provide benefit to the public may be confused with patent 'evergreening',[78] where otherwise obvious variations of prior inventions are granted patent protection inappropriately. Sometimes the distinction between incremental innovation and evergreening is a matter of opinion rather than fact.

The lines between the incremental innovation that merit patent protection and evergreening efforts that inappropriately exploit the patent systems and keep, for example, generic medicines from the public are often blurred. What can be said is that limiting protection for the incremental innovation that often derives from PROs may be detrimental to global innovation and access to medicines. For example, 'incremental innovations' that provide once-a-day dosing and acid stable antibiotics provide for greater patient compliance and accessibility. Similarly, heat-labile therapeutics support the ability to deliver temperature-sensitive drugs to markets lacking electricity and refrigeration. Thus, while some might call these types of innovations 'evergreening', they help to provide critical medicine technologies to populations that might not otherwise benefit from modern medical advances.

3.7 Patent Harmonization and Access to Medicines

As noted above, since the advent of the Bayh–Dole act in the United States, international attempts to emulate the US success and to harmonize patent standards

77) Straus, J. (2000) *Expert Opinion on the Introduction of a Grace Period in the European Patent Law Submitted upon request of the European Patent Organization.* European Patent Office, Munich. http://epo.org/about-us/press/releases/archive/2000/25072000.html.

78) 'Evergreening' is when patent owners attempt to extend the patent monopoly by seeking a new patent that 'updates' the first one before its expiration. This is usually done by claiming things such as an 'inventive' method for administering the pharmaceutical compound covered by the base patent. For pharmaceutical products, this means an extended monopoly that excludes generic drugs from the market.

has, by some estimations, resulted in greater patent rights, greater scope of exclusivity and decreased access to, for example, vital health technologies. According to Kapczynski *et al.*:[79]

> ... the United States, the European Union, and Japan have used trade agreements to impose high levels of substantive and procedural protection for IP on countries around the world. The World Trade Organization's Trade-Related Aspects of Intellectual Property Agreement is the foundation of this treaty architecture, but regional and bilateral agreements increasingly impose even higher protections upon countries ... This is particularly true in the area of medicines: at the time the Uruguay Round of trade negotiations was launched, more than fifty countries did not provide patent protection on medicines.

However, the establishment of the TRIPS Agreement provides that least-developed countries had until 1 January 2006 to comply with the terms of the agreement and have the right to defer patents and data exclusivity rights on pharmaceuticals until 2016.

Patent eligibility has played a significant role in the provision of technology, especially pharmaceuticals to the developing world. For example, India did not provide patent protection for pharmaceuticals until January 2005 when they became 'TRIPS compliant'. Before that time, India developed an extensive infrastructure based upon the manufacture of drugs that would have otherwise been patented. Indian companies continue to provide many low-cost drugs to developing countries. However, with the introduction of patent protection for pharmaceuticals, manufacturing of current generation drugs for delivery to developing markets has moved in significant part to other countries that are, in turn, developing manufacturing capacity. Thus, we are currently undergoing a 'TRIPS compliance cascade' that is helping with the establishment of manufacturing capacity throughout the world.

As this is not a treatise on international patent rights, a discussion of possible reasons for this cascade and its effects will not be discussed. However, what is clear is that the availability of patent protection and the scope of that protection has a significant impact on the availability of technology around the globe as well as the ability of countries to participate in this technological revolution. The ability for PROs and there faculty to participate in this revolution through the patent system has played a signification part in both the development and the deployment of technology. Appropriate safeguards that balance public and private interests is clearly the key maintaining the capital investment incentives provided by IP rights. However, there has been recent movement to dilute the strength and

79) Kapczynski, A., Chaifetz, S., Katz, Z. and Benkler, Y. (2005) Addressing global health inequities: an open licensing approach to university inventions. *Berkeley Technology Law Journal*, **20**, 1031.

vitality of the patent system and it is unclear as this point whether this will ultimately harm or help innovation.

Paragraph 5 of the Declaration on the TRIPS Agreement and Public Health adopted on 14 November 2001 by the WTO (the so-called Doha declaration) states in part that in cases of public health emergencies '... each Member has the right to grant compulsory licenses and the freedom to determine the grounds upon which such licences are granted'.[80] However, most nations reserve a royalty-free use license when they issue a patent. The harbinger of a nation being forced to license a technology carries with it a degree of unpredictability that haunts the business community. Such ambiguity may serve to undermine the incentives that patent exclusivities serve to provide.

The degree to which governments will employ the provisions of the Doha declaration remain to be seen. The United States has consistently resisted the use of compulsory licenses and other provisions of intellectual property law such as march-in rights[81] that would dilute the patent strength. Nonetheless, several countries, including Brazil, France and Ghana, have threatened to invoke the Doha declaration provisions for compulsory licenses for technologies that they felt were not being provided to the public at reasonable cost. Last-minute concessions by intellectual property holders have so far obviated the need for such licenses and therefore the impact of the Doha declaration provisions remain unclear.

3.8 Final Notes on the Global Expansion of Bayh–Dole-Type Intellectual Property Regimes

There is no universal panacea to control, regulate, and spur utilization of publicly funded technology. Mowery notes that '... indeed, emulation of Bayh–Dole actually could be counterproductive in other industrial economies, precisely because of the importance of other channels for technology transfer and exploitation by industry'.[82] What is clear though, is that the development of a flexible system that extracts and adapts the best practices of world intellectual property regimes and discards those that are not applicable within a particular country will ensure that an appropriate balance between public and private interests will be maintained. This balance is the key to providing continuing innovative activities that will guarantee that the innovation cycle will endure.

80) http://www.wto.org/english/thewto_e/minist_e/min01_e/mindecl_trips_e.pdf#search=%22wt%2Fmin(01)%2Fdec%2F2%22.

81) http://ott.od.nih.gov/policy/policies_and_guidelines.html.

82) See Footnote 60.

4
Current Intellectual Property Management Situation in Japan

Yukiko Nishimura and Katsuya Tamai

4.1
Introduction

Japan's postwar history highlights the national economy's domestic/international growth. Japan experienced postwar restoration, high economic growth centering on the manufacturing industry and the doldrums after the economic bubble burst from the late 1980s, suffering from the effects of the recession up to turn of the century. In the current century, Japan has continuously made efforts to enhance its competitiveness with new developments in the field of information technology and finance.

Under these circumstances, an intellectual property (IP) strategy – the strategic protection and use of research results and works as IP – is fundamental for the strengthening of Japanese industry's competitiveness and the revitalization of the national economy. The Japanese government has launched various projects and measures for the realization of an 'IP Nation'.

Universities have become more conscious of their new function – to make a social contribution. The incorporation of national universities in 2004 accelerated university–industry collaboration and technology transfer.

We overview IP-related projects for building up national power with the focuses on life sciences/biotechnology and university IP management systems.

4.2
IP-Related Government Measures and Projects

Since the enactment of the Basic Law on Science and Technology in 1995, the Japanese Government has poured a large amount of money into science and technology for the realization of a 'nation based on the creativity of science and technology'.[1] This Law stipulates the basic framework for science and technology

1) Basic Law on Science and Technology. http://www.mext.go.jp/b_menu/shingi/kagaku/kihonkei/honbun.htm (Japanese). http://www8.cao.go.jp/cstp/english/law/Law-1995.pdf (English).

Technology Transfer in Biotechnology. A Global Perspective.
Edited by Prabuddha Ganguli, Rita Khanna, and Ben Prickril

ISBN: 978-3-527-31645-8

policies to strongly promote science and technology by organized collaboration among the government, industries and universities.[2] Based on this law, the Basic Plan for Science and Technology was structured for the comprehensive and systematic implementation of projects for the promotion of science and technology.[3] This program focuses on specific fields such as life sciences, information and telecommunication, the environment, nanotechnology and materials, and pursues researcher training and active collaboration among the government, industries and universities.

Under this situation, the protection and utilization of the achievements of scientific and technological research was more emphasized for the recovery of the government's money allotment so far, which raised the public awareness of 'IP rights'. The Japanese government started the environmental improvement for the realization of 'IP Nation' as a next step for the 'nation based on the creativity of science and technology'. Table 4.1 illustrates the government's measures for IP management and industry–university collaboration.

Following the first policy speech made by Prime Minister Koizumi in February 2002, the government organized the 'Intellectual Property Strategy Council' to work out the 'Outline of Intellectual Property Strategy'. This outline is the government's basic plan for the realization of an 'IP Nation', clarifying the guidelines for each ministry's IP strategy focusing on four aspects, including IP creation, protection and utilization as well as human resource development. Further, in October 2002, the Intellectual Property Basic Act was enacted.[4] Article 2 of this law defines 'intellectual property' and 'intellectual property rights':

- 'Intellectual property' – inventions, utility models, new plant species, designs, works and other matters resulting from human creative activity (including discoveries or explanations about natural laws or phenomena, as far as they are susceptible to industrial application), trademarks, firm names and other means that are used in commerce to distinguish goods or services, and trade secrets and other technical or trade information that is important for business activity.
- 'Intellectual property rights' – patents, utility model rights, plant variety rights, design rights, copyrights, trademarks and other rights regulated by law or rights about interests protected by law.

2) Basic Law on Science and Technology, Article 2(2):

> In the promotion of S&T, the improvement of balanced ability of research and development (hereinafter referred to as "R&D") in various fields, harmonized development among basic research, applied research and development and organic cooperation of national research institutes, universities (including graduate schools in this law) and private sector, etc., should be considered, and in consideration of the fact that the mutual connection between natural science and the humanities is essential for the progress of S&T, attention should be paid to the balanced development of both.

3) Science and Technology Basic Plan (CABINET DECISION/July 2, 1996). http://www.mext.go.jp/b_menu/shingi/kagaku/kihonkei/honbun.htm (Japanese). http://www.mext.go.jp/english/kagaku/scienc03.htm (English).

4) Intellectual Property Basic Act (Act No. 122 of 2002). http://www.cas.go.jp/jp/seisaku/hourei/data/ipba.pdf (English).

Table 4.1 Policies relating to industry–university collaboration and the development of IP promotion in Japan.

Year of legislation or measure launched	Law	Measure	IP promotion
1995	Basic Law on Science and Technology		
1996–2000		Science and Technology Basic Plan (primary period)	
1998	Law Promoting Technology Transfer from Universities to Industry		
1999	Law on Special Measures for Industrial Revitalization Law for the Promotion of Start-up Enterprises		
2000	Law to Strengthen Industrial Technical Ability		
2001		Industrial Cluster Creation Plan	
2001–2005		Science and Technology Basic Plan (second period)	
2002	Intellectual Property Basic Act	Industrial Cluster Creation Plan	Biotechnology Policy Outline
2003	National University Corporation Law	University Headquarters Improvement Project	Establishment of IP Strategy Headquarters
			IP Promotion Plan
2004			IP Promotion Plan 2004
2005			Establishment of IP High Court IP Promotion Plan 2005
2006			IP Promotion Plan 2006
2006–2010		Basic Program for Science and Technology (third period)	
2007			IP Promotion Plan 2007

This clear-cut distinction between 'intellectual property' and 'intellectual property rights' legally established the solid concept that an achievement of science and technology research includes various 'intellectual properties', any of which under protection/regulation by law are to be 'intellectual property rights'. Further, this law stipulates (i) implementation of IP-focused measures; (ii) clarification of the responsibilities of the government, municipalities, universities and companies; (iii) planning and operation of the strategy for IP creation, protection and utilization; and (iv) organization of IP Strategy Headquarters.

After the enactment of Intellectual Property Basic Act in March 2003, the cabinet established the 'IP Strategy Headquarters' for the intensive and systematic implementation of the relevant measures based on the law. The Headquarters comprises 28 members including the Prime Minister as the chief, all the ministers and experts in the private sector. Its main functions are the establishment of an IP strategy and the promotion of its implementation, as well as any arrangements of the IP-related policies.

At the same time, the Cabinet Secretariat established the Secretariat of the IP strategy Headquarters. This Secretariat supports the Headquarters' smooth proceedings and arranges the administrative work for the above IP policies. Mr Toshimitsu Arai, the former representative of Intellectual Property National Strategy Forum, was appointed as Director-General.

The Headquarters issued the 'Promotion Plan for IP Creation, Protection and Utilization of IP' (the 'Promotion Plan'), which is a so-called protocol for the Japanese economy's transformation into an 'IP Nation' under the basic concept of the Intellectual Property Basic Act. The Promotion Plan has the following three policies:

1. To create unconventional exceptions for IP.
2. To establish an internationally competitive system.
3. To realize timely and rapid reform.

Under these policies, the IP Strategy Headquarters held five consecutive meetings to approve the Promotion Plan with 270 items, which extensively and minutely describe Japan's ideal IP-related policies.

After the approval of the Promotion Plan, the government, each ministry and municipality took the first step for the realization of an 'IP Nation' to implement the specific measures according to the Plan. The rapid arrangement of this Promotion Plan in 6 months shows that the creation of the protocol for an 'IP Nation' was the government's pressing issue at that time.

The Promotion Plan is renewed annually every June. The plan in 2006 provides the basic guideline for a 3-year IP strategy till 2008 in pursuit of 'the most advanced IP Nation in the world', in which 600 items are proposed for the enhancement of international competitiveness by the uses of IP and related system development. Specifically speaking:

1. IP creation
   - Integration of university IP Headquarters and the Technology Licensing Organization (TLO), and strengthening of their affiliation.
   - Improvement of an integrated information search system for patents and research papers.
2. IP protection
   - Promotion of the Headquarters' effort for rapid and efficient patent examination.
   - System development for preventing leakage of technology by patent application.
   - Realization of patent's integrated approval by the patent offices in the United States, European Union and Japan.
   - Early establishment of a non-proliferation treaty for counterfeit goods and pirates.
   - More ridged regulation for private imports, etc.
3. IP utilization
   - Promotion of international standardization.
   - Support for small and medium-sized ventures.
   - Regional development utilizing IP.
4. Creation of a cultural nation making the most of content
   - Realization of a content superpower.
   - Structure of branding strategy.
5. Personnel development and public awareness enhancement
   - Implementation of an integral strategy for IP personnel development.
   - Training of international IP experts.

The 3-year IP strategy starting from 2006 is regarded as the second stage for the world's most advanced 'IP Nation', which emphasizes (i) more effective measures for an 'IP Nation', (ii) IP utilization for stronger international competitiveness and (iii) system design to cope with new challenges. Further, in the field of life science and biotechnology, there have been reviews and discussions in comparison with the overseas situation regarding (i) the study of research tool patents, (ii) protection of advanced technology under the patent system and (iii) securing experts for technology transfer, etc.

4.3 Life Sciences/Biotechnology-Related Projects

Japan fell far behind the United States and other countries in R&D in the field of biotechnology. In the United States, the President took the leadership role for the promotion of life sciences research as a top priority to inject approximately US$ 27.3 billion (3.3 trillion yen) as its research budget. On the other hand, the Japanese government allocated approximately US$ 3.7 billion (440 billion yen) (as of 2002) for research and development in the field of life science, equivalent to less than one-seventh of the US budget.

As for the shares by country of international patent applications related to biotechnology, the United States accounted for 52%, European Union countries 21%, and Japan 20%.[5] Under the circumstance where the number of patent applications from China has been rapidly increasing, Japan urgently needed to take specific countermeasures. Further, the number of Japanese university graduates in biotechnology (biotechnology, pharmacy, medicine, agricultural chemistry) in 1998 was 20 987 students with the BS degree, 3154 students with the MS degree and 3373 students with the PhD degree. As for the United States, these figures for university graduates in the biological sciences were 67 112 students with the BS degree, 6368 students with the MS degree and 5854 students with PhD degree. Compared with the situation in the United States, it is clear that the supply of manpower in this field at that time was not always sufficient in Japan.[6]

As of 2002, the accumulated number of start-up companies indispensable for biotechnology-related industries was 300 companies, much less than the number in the United States. Japanese venture capital invested more than US$ 45 million (5.4 billion yen) in total in biotechnology start-up companies, most of which were overseas companies, which resulted in a drain of capital from Japan.[7]

When it comes to state-of-the-art medical technology, Japan is under severe conditions. For pharmaceuticals, new medicines in the domestic market are mostly foreign products. Japanese pharmaceutical company's exports are half the amount of the imports. The number of medical patents obtained by Japan is merely one-fifth of the number obtained by the United States. According to the trends in new drug discovery and development (1996–2001) by the Ministry of Health and Welfare, 144 new components were developed for foreign products, compared to 66 components for domestic products. A company's R&D expenditure is 199 million USD (231 billion JPY) in the United States, compared to 40.3 million USD (48.8 billion JPY) (Source: the Japan Pharmaceutical Manufacturers Association (JPMA), 2000). As for the number of patents obtained by the United States and Japan in 3 areas of the world including Japan, the United States and the EU, there were 4030 patents obtained by the United States, compared to 1538 by Japan (Data from WIPO and USPTO, 1999), according to the Secretariat's Intellectual Property Strategy. Medical devices are also mainly from overseas and the number of patents related to medical devices owned by Japan is about one-seventh of the United States. The improvement of the Japanese medical level to meet the expectations of the people and patients by strengthening medical technology is one of the most significant issues for the nation.

Under such a situation, the Japanese government recognized the significance of the establishment of a national strategy for life sciences and biotechnology on a grand scale. In 1999, five ministries relating to biotechnology (Ministry of International Trade and Industry, Science and Technology Agency, Ministry of Health and Welfare, Ministry of Education, Culture, Sports, Science and Technology and Ministry of Agriculture, Forestry and Fisheries at that time) structured the 'Basic

5) Researched by the Japan Patent Office; covering patent applications from 1990 to 1998.

6) Basic Survey of School, MEXT.

7) Biotechnology Strategy Council. http://www.kantei.go.jp/jp/singi/bt/kettei/021206/taikou.htm (Japanese)

Strategy for the Creation of Biotechnology Industry'. Based on these strategies, the 'National Strategy for Industrial Technology (Biotechnology)' was established followed by the 'Millennium Project' including all the basic policies and plans. This project's goal is to create government–industry–university collaboration projects focusing on technology innovation by concentrating the researchers' intellectual abilities under the government initiative where five research projects (i.e. human genome multiparametric study; affected gene; bioinformatics; generation, differentiation and renaturation; and rice genome) are launched under the three significant and pressing themes for Japan's future economic society (i.e. information technology, aging and environmental responsiveness). Such circumstances forced Japan to build up a national strategy to strengthen and promote the field of biotechnology.

In December 2002, the 'Biotechnology Strategy Council', consisting of the Prime Minister, the ministers of the related ministries and experts, produced the 'Outline of Biotechnology Strategy'. This outline stresses that the development of 'life sciences', especially 'biotechnology', will influence the fate of human life in the twenty-first century, and that social consensus on the safety and ethical issues is indispensable for the development of biotechnology. It further states that any intensive enhancement measures should be taken in this decade, considering that rapidly advancing biotechnology was becoming substantially emphasized on a global scale. It provides three strategies ('unparalleled improvement of R&D', 'drastic enhancement of industrialization process' and 'promotion of people's deep understandings') to realize 'good health and longevity', 'safer and more functional food' and 'sustainable development for comfortable environment', global contribution in the field of biology, strengthening of the global competitiveness of Japanese industries and new business creation.

The following are the goals relating to IP:

1. Clarification of handling of medical patents/patent for the cubic structure of proteins.
2. Invention remuneration system (based on university incorporation).
3. Review of arbitrary license system (for the case of research tools, technical standard).
4. Shortening of examination period of new varieties.
5. Strategic IP management at IP Headquarters, TLOs.
6. Establishment of Intellectual cluster, industrial cluster.
7. Personnel development (Management of Technology, etc.).

4.4 Medical Patent/Patentability

Apart from the preparation for the R&D environment, including the research foundations or various systems such as pharmaceutical regulations or insurance systems, the patent system intended for the promotion or diffusion of technological development can also be one of the most significant issues in the encouragement of the development of advanced medical technology.[8]

Inventions which could be under patent protection are defined as 'highly advanced technical ideas or creation developed by use of natural laws' and classified into two types: 'invention of a method' and 'invention of a material'. The former type is further divided into 'invention of a method to produce a material' and 'invention of a mere method'. Medical devices and pharmaceutical products are considered 'inventions of materials' by themselves and the methods to produce these products are 'inventions of methods to produce materials'; therefore, there could be subjected to patents.

The 'methods' associated with 'surgical operations, treatments or diagnosis of humans' had not been subject of patents previously in Japan according to the operational criteria described in the Patent Law, Article 29 ('no possibility for industrial use').[9] This is based on the idea that a medical treatment is not regarded as an 'industry'. Furthermore, due to a policy-based reason it was thought that the patent system is not an incentive since inventions of medical treatments take place in universities or university hospitals and medical research is not meant for R&D competitions, and due to humanitarian reason that it is not reasonable that the provision of medical treatments, which requires more emergency acts than the use of pharmaceutical products or medical devices, would not occur without a permission from the patentees.[10] Therefore, not only the skills involved in the doctor's actions, such as surgical operations, medical treatments or diagnostic methods, but also the technologies related to medical devices or methods of treatments had all been considered as the 'methods to operate, treat and diagnose humans', and hence excluded as patent candidates.

In recent years, new fields, such as gene therapy or regenerative medicine, which do not fit in the traditional concepts of medical treatments, have been as greatly developed as life sciences or biotechnology. In these fields, it is very common that the 'methods' of using identified genes or cells become the bases of inventions. However, no matter how sophisticated 'the methods of inventions' could be, these were excluded from the patent candidates as long as they are considered as 'methods to operate, treat and diagnose humans' (i.e. medical treatments). Thus, it was not possible to gain protection under the patent law until these were patented as 'matters' or 'methods to produce matters'. In addition, in order to substantiate these as 'matters', a huge amount of investment for the secure safety as well as high technology/capital investment offered by companies are required; hence, there is more need for right protection. Thus, the issue has been raised if the right protection is necessary even at the stage of 'methods'.[11]

8) Industrial Policy Coordination Meeting for Pharmaceutical Products and Medical Devices, Pharmaceutical Industrial Vision, 2003 and 2004.
9) Review Criteria for Patent/New Practical Ideas, Section II, Chapter I.
10) Handling of Medical Related Activities in the Patent Law, Sub-committee for the Patent Law of Intellectual Property Division of Industrial Structure Review Board, 2003.
11) See the Reference. Tsujimaru, K. (2003) Medical Treatment and Patent. *J. Information Processing and Management*, **46**, 107–112 (Japanese).

In the United States, ever since the Patent Act was amended in 1952, all methods that are broadly related to medical fields have been the subject of patents and numerous medical patents have been established. In 1996, an amendment to the Patent Act was raised based on an issue regarding *Pallin v. Singer*,[12] which occurred in 1993. Then, a regulation was introduced that patents are not principally valid in the medical treatments provided by physicians, etc., while patents for medical methods are still active.[13]

In Europe, the European Patent Convention was concluded in 1973 and it was provided in the treaty that treatment/diagnostic methods in surgical operations or clinical trials with humans or animals were not to be subject to patents. Hence, surgical operations, treatments and diagnostic methods are not considered as patent candidates. However, as an exception, diagnostic methods which are limited to data collections or comparisons are not considered as diagnostic methods and therefore subject of patents.[14] Also in Europe, there is a recent movement to protect medical/pharmaceutical inventions by accepting that, even in a case where it was characteristic only in dosage or indications, it is the 'use of materials for productions of pharmaceutical products necessary for certain treatments which are new and progressed' as a Swiss-type claim.[15]

Thus, in the Patent Act, the handling of actions associated with medical treatments is categorized into 'upstream regulations' and 'downstream regulations'.[16] Depending on these two different regulations, inventions related to medical practice are treated differently at the patent examination. Under the 'upstream regulation' system, a medical practice-related invention is excluded from patent examination; under the 'downstream regulation' system, the invention is subject to patent examination. The 'upstream regulation' indicates a position in which an invention involving 'medical practice' is to be excluded from the patent candidates at the time of patent review. However, the 'downstream regulation' indicates a position in which the invention is to be included as a patent candidate at the time of patient review although a medical practice performed by a doctor, etc., is considered as a collateral issue, at the time of patent activation, by establishing the exceptive orders for doctors and medical institutions.

In Japan and Europe, where the former type of regulation has been established, patentability criteria are released from the patent office as a guideline to figure out if the action is 'medical practice' or not. However, it is problematic in terms

12) *Pallin v. Singer*, 36, USPQ 2d 1050 (Va, 1995).
13) Refer to US Patent Law [35 USC §287(c)(1)]. However, in the case where (i) patents for materials such as medical devices or pharmaceutical products, (ii) patents for the method of using pharmaceutical products, (iii) patents for biotechnology were violated, the patent owner is able to exercise his/her right against doctors or medical institutions as well [35 USC §287(c)(2)(A)].
14) Upon the amendment of the European Patent Convention in 2000, introduction of patents for medical treatment methods was discussed, although it did not reach an agreement. At this time, the official reason for not patenting medical treatment methods was changed from 'not considered as inventions for the industrial use' to 'excluded from patentable candidates' to make it consistent with the TRIPS Agreement. See Footnote 7.
15) European Patent Office Decision T-1020/03.
16) Koizumi, N. (2002) Legal Policy for the Protection of Invention Relating to Treatment Method. *Jurist*, **1227**, 40–6.

Table 4.2 Comparison of medical activity patentability in the United States, Europe and Asian countries [adapted from Tamia, K., Kaneshiro, K. and Nishimura, Y. (2005) Gordon Research Conference].

	Location	What provision is it based on?	How is medical activity delimited/
Upstream regulation (patent examination)	Japan, South Korea, China	Not industrially applicable invention [Japanese Patent Law 29(1), South Korean Patent Law 29(1), Chinese Patent Law 25(4)	List of case examples (examination guidelines)
	EurPat	Not susceptible to industrial application [European Patent Convention 52(4)]	Build-up of decisions (examination guidelines)
	EurPat (revised 2000)	Exceptions to patentability [European Patent Convention 53(c), revision]	Build-up of decisions (examination guidelines)
Upstream regulation (patent enforcement)	United States	Limitation on damages and other remedies [35 USC §287(c)(1)]	Provision in the Patent Act [35 USC §287(c)(2)(A,B,C)

of the stability of law since the criteria needs to be amended repeatedly whenever a new technology is introduced. On the other hand, the United States takes the latter position. A comparison of medical activity patentability issues in the United States, Europe and Asian countries (in 2004) is shown in Table 4.2.

Namely, it was found that the present system in Japan provides a narrower range of patent protection compared to the United States or Europe. Therefore, venture business involving medical activities had been intending to recover the capital by obtaining patents for the technologies around the medical activities. The current Japanese system adopts a limited scope of patent protection compared to the United States and European Union. Many venture companies relating to medical practice are handling medical practice-related technologies such as regeneration medicine. For them, obtaining patents is vital to compete with large companies with sufficient funds. However, they cannot obtain any patents for a medical practice itself. Under this situation, they tried capital recovery by obtaining patents for the medical practice's peripheral technologies. When they cannot expect a medical practice-related technology as a means of their capital recovery, a venture company's medical practices cannot be promoted, or even developed.

Therefore, many discussions have been held so far regarding the protection of medical activities by patent law. The necessity of clarification with the handling of technologies associated with regenerative medicine or gene therapy in patent law was indicated in the 'Outline of Intellectual Property Strategy', which has been previously described. As a consequence, the patentability criteria were partially amended in August 2003 and clearly specified a method 'to manufacture' pharmaceutical products or medical devices using materials derived from humans such as a bioengineered skin product used for regenerative medicine, etc., that it is subjected to a patent even if the product is conditioned to be returned to the same person (autologous transplant).[17]

Also in 2005, the amendment review criteria (proposal) for 'inventions which can be industrially used' was released and it was clearly specified that 'an operation method of a medical device' can be considered as a description of a function that the medical device has as a method and, hence, it can be patentable. It was also noted that a method that includes actions made by an operator (e.g. actions made by a doctor in order to operate a device) or a function of the device on human bodies (e.g. resection of a lesion using a device) cannot be included as a patent candidate. Thereafter, amendments and additions of issues were made in the review criteria in 'a proposal of a patent for the practical use regarding 'a method to operate, treat or diagnose a human'.[18] In other words, it was concluded that 'an operational method of a medical device' should be patented generally in a condition such that skills associated with the physician's practice are not subject to patent protection and thus patent protection came to be possible with the method of invention by a method claim.[19]

On the other hand, in terms of patent protection regarding 'a method to obtain a report of new indications of the pharmaceutical products for manufacturing/marketing these products', the operation of patentability review was further clarified in 2004 in descriptions, new/developing qualities, etc., of the medical/pharmaceutical inventions, mainly relating to the issues that require particular judgments/handling. In addition, for medical/pharmaceutical inventions that can be specified by the combination of multiple products or treatment styles such as dosage/interval administrations, it was clearly stated that these were 'inventions of matters' and therefore should be handled as 'inventions for industrial use', and the evaluation methods for patentability in terms of new/developing qualities, etc., were also clarified.[20]

For patent protection of a 'method to cause a pharmaceutical's new effect and efficacy for its manufacturing and marketing', a more clear-cut patent examination system for the applications that need more specific consideration or treatment for a medical invention's description or its novelty/inventive step, etc., was structured in 2004. In addition, the medical invention specified by the combination of several medicines or treatment approaches such as dosing interval and dosage, which is regarded as 'an invention of a product', is to be expressly

17) Kaneshiro, K., Masuda, S., Tanaka, Y. and Tamai, K. (2004) Protecting methods for treatment related to regenerative medicine and gene therapy in Japan. *Nature Biotechnology*, **22**, 343–345.

18) 'Inventions for the industrial use' 2.1.1. (1); Review Criteria for Patent/New Practical Ideas, Section II, Chapter I.

19) Biotechnology Committee; Protection for Medical Related Activities. *Intellectual Property Control*, **55**(1), 41–8.

20) Japan Patent Office website. www.jpo.go.jp.

handled as the 'inventions for industrial use'. The evaluation method of its patentability including novelty and inventive step, etc., was also clarified.

Thus, almost all the categories/elements for a 'operation method of a medical device' can be included in its claim under the situation where the description of a method in addition to a product is permitted. In this way, all the elements of an invention became patentable by this amendment of the patent examination standard.[21)]

In the future, it will be necessary to discuss 'treatment methods' while clarifying that, with medical activities, the patent is beyond the valid range. At this time, attention is required not to be confused since the line to differentiate the activities related to the medical treatments from the medical practice may vary by country.[22)] Moreover, it is considered that there is a limit to the method where the method claim is being simply added to the corresponding review criteria in order to protect the 'invention of the method', including the 'treatment method', as the 'invention of the matter'. The protection of advanced medical technology will be further required in the future for the improvement of medical services in Japan and for winning the international competition. Therefore, more drastic designs will be necessary in the system, considering the amendment of the patent law and keeping up with the other countries in the world.

4.5 Policies related to University–Industry Collaboration – Incorporation of National Universities

The biggest changes seen in universities in Japan these days are the collaborations between university and industry using the intellectual achievements from university, activation of technology transfer and incorporation of national universities. Since national universities used to be under the total control of the former Ministry of Education in Japan, it used to be very difficult for the universities to receive research funds offered by other Government offices. Therefore, most of the practical public funds were either used at research organizations in each Government office or in the public sectors, or allocated to the 'three public companies', especially, Nippon Telegraph and Telephone Public Corporation or Japan National Railway Companies.[23)] However, ever since the Primary Basic Program for Science and Technology was introduced in 1996, as mentioned previously, huge amounts of public funds started to be invested in universities.

21) See above Footnote 19.

22) For instance, regenerative medicine is considered mere material once it is detached from the skin; hence, the method to remove this is not considered as a medical activity in Europe. On the other hand, 'medical activity' includes an activity that is defined as a medical practice in Japan (an action which might induce a risk to the human body unless otherwise performed by physicians' with his/her medical judgment and skills).

23) Tamai, K. (2005) Industrial liaison and university identities. *Japanese Journal of Applied Physics*, **75**, 63–8.

In 1998, the TLO law,[24] 'the law regarding the transfer of technological research achievement from universities, etc., to the public business', was established and the preparation for the practical system of industrial liaison, including the approved TLO, was started in local settings.[25]

The TLO is an organization established mainly for the purpose of patenting inventions developed by university faculty members, searching for companies to be licensed and obtaining the licensing fees by making the licensing contract with them. The profit is partly returned to the researchers and turned into further research funds, contributing to further activate academic research. Hence, it was expected to become the basis of the 'intellectual creation cycle'. A subsidy, up to a maximum of US$ 250 thousand (30 million Japanese yen), is granted for up to 5 years for the TLO approved by the Ministry of Education, Culture, Sports, Science and Technology (MEXT) and the Ministry of Economy, Trade and Industry (METI) [although the subsidy rate is within two-thirds; the company needs to raise US$ 83 thousand (10 million yen) for US$ 167 thousand (20 million yen) subsidy].

However, the national universities had not been incorporated yet at this time; hence, they had not been allowed to possess any patents. Therefore, in many cases, the faculty members voluntarily established the TLO as a stock company outside the university territory and held the stock. In December of the same year, three TLOs (Tohoku Techno Arch, Ltd, established by some faculty members from universities including Tohoku University; Toudai TLO (CASTI), Ltd, established by some faculty members of the University of Tokyo; and Kansai TLO Co., Ltd, established by faculty members of universities such as Kyoto University and Ritsumeikan University) were approved. The number of approved TLOs increased thereafter and there are 42 organizations all over the country at the present time. Most of them are still in the form of private companies.[26] There are also four approved TLOs that are allowed to deal with government-owned patents held by national research organizations as a business. However, there is no need to become an approved TLO if they are not requesting any subsidy granted by the

24) TLO (Technology Licensing Organization) is used as a general term in Japan since it had been used to indicate a Technology Transfer System according to Stanford University and Massachusetts Institute of Technology. However, in the United States, OTT (Office of Technology Transfer) or TTO (Technology Transfer Office) is more generally used to indicate a Technology Transfer System in universities rather than TLO. OTM (Office of Technology Management) or PTM (Patent and Technology Management) are also used. Unlike Japan, a TLO in the United States is usually a single on-campus organization and there is hardly any off-campus organization. It will be further described later. In Europe, there are many terms using 'Technology Transfer' or 'Innovations'. Not only on-campus organizations, but also profit-making off-campus organizations exist in Europe. The details will not be discussed in this chapter, but are discussed elsewhere in this volume.

25) The law regarding the transfer of technological research achievements from universities, etc., to the public sector. http://www.mext.go.jp/a_menu/shinkou/sangaku/sangakuc/sangakuc10_1.htm.

26) Out of 42 organizations, 23 are private companies, 10 are foundations and nine are on-campus organizations.

Government. Hence, there are many small private universities which run their own TLOs.

The Industrial Revitalization Law (revised in 2003), which was established in 1999, made it possible that the IP rights in a business consigned by the government revert not to the government, but to the consigned company, universities or researchers.[27] This law includes a section called the 'Japanese Bayh–Dole Act', which was modeled on the Bayh–Dole Act in the United States.[28] This law enabled the IP right with the result obtained in the R&D consigned by the government to belong to the consigned company for 100%. Moreover, a policy that is equivalent to the Bayh–Dole Act in the United States, stating that research achievements belong to the researcher even if they were funded by the government, was introduced and this further enabled organizations to possess IP rights that used to be held by national universities, after incorporation of these universities. The Law for Facilitating the Creation of New Business Creation was established in the same year in order to provide more support to venture companies and it greatly helped university-initiated ventures as well.

In the Law for the Enforcement of Industrial Technology, which was enacted in 2000, faculty members of national universities came to be allowed to simultaneously take positions outside universities, such as being a research instructor for a company or an executive member of a profit-making company. The latter position was especially appreciated in the promotion for start-up venture businesses based on the technological skills that the member can provide. The approved TLO was allowed to use the facilities of national universities as well without any charge. Furthermore, in 2002, the teaching faculty working for national universities came to be allowed to have side jobs, as executive members of either TLOs or companies that utilize the outcome achieved in the research, not by the MEXT but by the dean of the university. In 2003, the teaching faculties working for the national universities were allowed to do the outside jobs even during university office hours.[29]

As a major change following this, The National University Corporation Law (2003), which accepts the position of a national university as a corporation and appreciates its independency, was established while the MEXT showed some tendency to enforce the control. In April 2004, 99 national universities and 15 organizations collaborating with universities were renewed as 89 national university corporations and four corporate organizations collaborating with universities. It was the beginning of the change from the previous researcher (laboratory)–company relationship to the university–organization relationship.

Until then, they had been regarded as part of the government. Now that their management system has become underpinned by the idea of autonomy and self-

27) Industrial Revitalization Law. http://law.e-gov.go.jp/htmldata/H11/H11HO131.html (Japanese).

28) The Japanese Bayh–Dole Act represents the Industrial Revitalization Law, Article 31 (handling of patents associated with achievement in researches consigned by the Government).

29) They are allowed to work in a non-executive capacity on the side everywhere, but in an executive capacity only in the special area designed for structural reformation.

responsibility, they are expected to be more efficient. In addition to the annual evaluation of the performance held at each university, their management efficiency is evaluated every 6 years.

The incorporation triggered a major change. One of the most drastic changes they experienced is the ownership of IP. Figure 4.1 shows a flow diagram of intellectual achievements at Japanese national universities before/after incorporation.

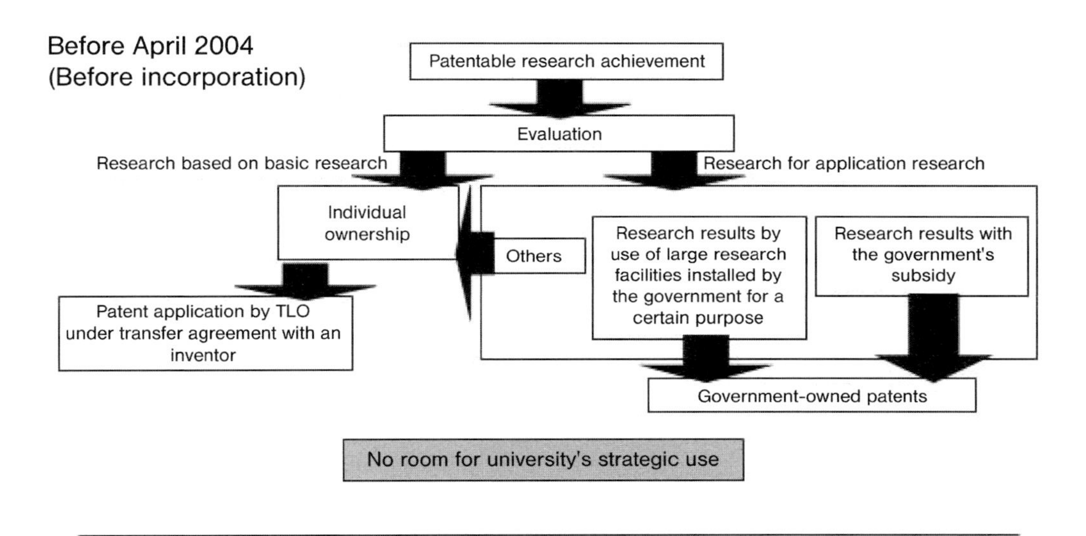

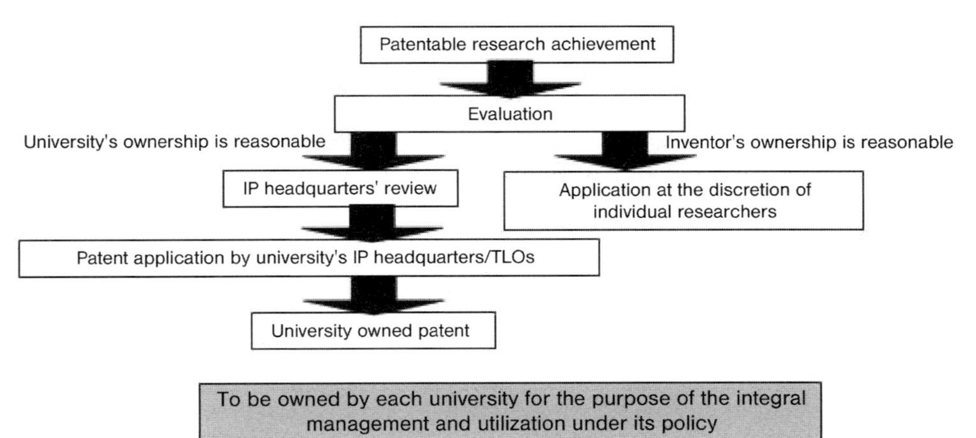

Figure 4.1 Changes of IP management systems before/after incorporation.

In fiscal year 2002, Japanese universities had 3832 inventions in total. Among them, 682 inventions were government owned (17.8%) and 3150 inventions (82.2%) belonged to the inventors themselves.

Since that time, some intellectual achievements such as inventions originally owned by the inventors were patented by TLOs, based on the assignment agreements allowing the inventions to be eventually transferred to the relevant industries. The inventions not owned by the government were patented by the companies as research partners.

Under these circumstances, universities did not realize what was happening with the inventions and patents obtained as the outcome of their faculties' efforts nor could they utilize such patents strategically at their own discretion.

The National University Corporation Law provides that the national universities shall promote the spread and utilization of their research results.[30] This law clarifies the university's significant responsibility to create intellectual achievements and to focus on making the most of such achievements. Concerning the ownership of IP at universities, MEXT encourages the universities to handle and utilize their IP according to their own policies, following the principle of institutional ownership – in view of the conventional insufficient utilization of intellectual achievements under inventors' ownership, the recent development of the system and the incorporation of the national universities.[31]

The preparation for the system proceeded. In July 2003, MEXT determined the implementation of the 'University's IP Headquarters Improvement Project' with 33 universities and one institute. The aim of this project was to promote the universities' development in a way that they contribute to society by the use of their IP.[32]

The duration of the project is principally 5 years. More precisely, the following objectives were raised: (i) establishment of clear rules including a poly for intellectual properties, etc., (ii) preparation for the systematic management system involving the entire school, (iii) meeting the needs of human resources (from outside) specialized in the intellectual properties, and (iv) enforcement of an effective and efficient utilization system. As seen with TLOs, there are some universities that established their own IP law centers other than the centers listed here. According to their advice, universities all around the country prepared the guidelines for the policy of IP or the policy for the conflict of interest as well as the industrial liaison guidelines. Under such policies and measures, the basis of the intellectual achievements had been the principle of institutional ownership after the university's incorporation.

30) Article 22, Paragraph 1, Item (5) of the National University Corporation Law (Law no. 112, promulgated on 16 July 2005, enacted on 1 October 2005).

31) MEXT Council for Science and Technology, Technology, Division of Technology and Research Infrastructure, Panel for the Promotion of Industry–University–Government Affiliation (2003) *Structure of Industry–University–Government Affiliation for a New Era (Discussion Report)*, Chapter 5. Systematic Management and Utilization of Intellectual Property at Universities, etc.

32) Announced by MEXT on 15 July 2003. In addition to the 'University's IP Headquarters Improvement Project', nine organizations are adopted for 'Support Program of Distinctive Functions for IP management and Utilization'.

Consequently, the number of universities' intellectual achievements smoothly increased after the incorporation. If we compare the incorporated national universities' industry–university collaboration and technology transfer in 2005 to the performance in 2001, it increased by 2.5-fold in the number of inventions, 15.5-fold in the number of patent applications and 15.8-fold in the number of licensing agreements, and the licensing profit has more than doubled.[33)]

Thus, industrial liaison/technology transfer laws have been established amazingly rapidly in Japan since the late 1990s. However, it is not the case that the liaison between universities and public companies also started suddenly in the late 1990s. Even before this time, collaborative research with public companies or the transfer of research outcomes (including the patents) from universities, etc., to the public took place on a small scale or sometimes within a direction of an individual teaching faculty.

The question is why this big movement of industrial liaison occurred during this specific period. It used to be considered that the establishment of an 'innovation system' was urgently needed as a solution against the decrease in industrial competitiveness in Japan, which occurred due to the prolonged economic recession, and to fulfill the need to enforce competitiveness. The 'innovation system' stands for the system of the entire country to stimulate innovative actions (technological revolutions).[34)] It includes not only the development of new products or introduction of new manufacturing technologies in a company, but also the broad concepts of providing adequate capital to encourage more innovative creations or a policy regarding intellectual properties. Therefore, in addition to public companies, universities or public research organizations are also included in the sectors associated with this corresponding system. It can be said that the movement has been greatly promoted until now based on the discussion of how to collaborate productively the intellectual achievements (innovations) obtained in these organizations (i.e. how to conduct the industrial liaison/technological transfer effectively) on a large scale.

Moreover, this flow dramatically changed the notion of research and the society that had been possessed by the researchers working at universities, etc., or the universities themselves. Due to the introduction of various policies that had been conducted since the late 1990s, many researchers at universities who had been locked in the 'ivory tower' were forced to come out of the socially isolated position and face the outside world. As a consequence, various collaborative works with the outer society were activated and the research ability of the researchers was further extended. On the other hand, there are problems caused by such a dramatic change.[35)]

33) Nishimura, Y. (2007) The current status of handling, of students intellectual accomplishments in Japan. *IPR Info*, 1, 30–32.

34) Nelson, R. (1993) *National Innovation Systems: A Comparative Analysis*, Oxford University Press, New York. OECD (1999) *Managing National Innovation Systems*, OECD, Paris.

35) Nishimura, Y. (2007) Industrial liaison and technology transfer management in Japanese universities, in *Industrial Liaison in Japan* (eds Y. Miyata and K. Tamai), Tamagawa University Press, Tamagawa, chapter III, 81–118.

There are various forms of industrial liaison/technology transfer, and it is obvious that the meanings of 'transfer' and 'liaison' can be very broad. At the present time, universities and public companies are able to select the form that can produce benefits for both, or even for the society, out of the many choices.

In order to conduct these industrial liaison/technology transfer activities, the effort made by researchers, who take a role of creating intellectual properties, is not enough, of course. As described above, industrial liaison/technology transfer-related policies have been established one after another, and intellectual achievement management organizations and associated institutions in national universities, etc., have also been maintained without major problems.

However, the discussion regarding the necessity of handling university-born intellectual achievement, which involves the university or the community, is still continuing. There are still some researchers who consider it is not their job to think about IP (management) since 'IP management = patent application control'.

Moreover, over-establishment of policies resulted in the extremely complicated situation in which both the TLO, which mainly exists as a private company or a public corporation, and the IP office, which is recognized as an on-campus organization, perform the management at the same time. They merely reflect the outcomes of the measures taken by each setting from the aspect of utilization of pre-existing organizations or subsidies, and therefore, it is not the system which it was expected to be in 1998.[36] As a result, various relationships were created in terms of coexistence of the TLO and IP center, depending on the situations for each university or the community. Some people have brought the discussion back regarding the significance of TLO which has not been in black. The complexity of these relationships sometimes confuses university faculties, the inventors and the companies that are utilizing their achievement, eventually inducing a stagnation of the management.

Industrial liaison/technology transfer (i.e. the management of intellectual properties) is to manage the 'intellectual properties' of researchers. It includes two types of management systems: the 'management of intellectual achievements which (might) be useful in the market at the present time' and the 'management of intellectual achievements which would (might) contribute socially in the future'. As far as time is concerned, the former can be considered as the 'narrow' definition of industrial liaison/technology transfer and the latter as the 'broad' definition.

In terms of the 'narrow' definition, it is easy to estimate the profitability since the product can appear on the market in a short period of time. However, it is not that the intellectual achievement is introduced from the right to left. Yoneyama *et al.* have noted that 'details of technology reported by an inventor, described in the submitted patent specification and transferred to a company are all differ-

36) Watanabe, T. (2005) *Discussion Regarding TLO and Administrative Duties of Intellectual Property Center.* Material distributed in the Sub-committee for Promotion of Industrial Liaison in Industrial Structure Council Industrial Science Technology Policy Committee.

ent'.[37] Furthermore, after the technology transfer (i.e. after introducing its utility in the market) there are many continuous administrative duties to be performed for the adequate use of the intellectual achievement.

As for the 'broad' definition, a tremendous amount of time is required until it actually contributes to society since the intellectual achievement is not matured enough. In addition, the utilization method needs to be decided out of many choices. Therefore, it seems difficult to consider the profitability where the observation is necessary in the long-term perspective. The meanings of 'social contribution' and 'utilization' are also different according to each university or community as well as to the types of intellectual properties possessed (or could be created) by them.

However, whether it is 'narrow' or 'broad', it is obvious that the intellectual profit management organizations do not only transfer the (highly patentable) technology to the industrial world from universities, but that they also play a part of a process that further refines the university-born intellectual achievement. Various strategies and cooperative relationships with researchers are necessary for this. In addition, introduction of the achievement to the social market, regardless of the duration, can further stimulate research activity and possibly open the door to the next research.

Professor Nelson of Massachusetts Institute of Technology has raised four objectives for industrial liaison/technology licensing in universities: (i) creation of employment opportunities, (ii) maintenance of technological competitiveness, (iii) encouragement of creation of socially revolutionary technology and (iv) financial support of universities from the royalty income (licensing fees).[38] All of these are significant objectives in Japan as well.

In Japan, the focused tends to be that the IP management organizations are established in expectation of achieving objective (iv). However, the royalty income is only 3–4% of the entire research expense in the United States as well. In addition, it is rather an accidental profit most of the time and, hence, is not necessarily attributed to the business strategies.

Other than this, a new evaluation basis is necessary to figure out if it has contributed to the local economy. Unlike private companies that can freely move their factories or laboratories according to business strategy, universities are characterized by the fact that they cannot move easily. In other words, universities need to settle deeper in the local community than private companies. It is necessary to create universities that concern the local communities from both the research (the creation of intellectual achievement) and educational (the cultivation of human resources) aspects especially after the incorporation.

Furthermore, local government cannot move from the community either. Namely, the university and the local government would likely to become the key players in the local economy. The example of successful industrial development

37) Yoneyama, S., Fukushima, M., Senoo, D. and Watanabe, T. (2006) Marketing of technological knowledge: empirical analysis of licensing activities from university TLOs to industrial sectors in Japan. *Technology Management for the Global Future, PICMET*, 4, 1865–74.

38) See Footnote 34.

based on the university-born intellectual achievements is seen in Silicon Valley and the example of company activations by providing the university-born intellectual achievements to the community is seen in the local clusters (or science parks) which have been developed in some states such as North Carolina. The communities can receive tax incomes from venture companies that have been using the university-born intellectual achievements or local industries.

From this point of view, in terms of industrial liaison/technology licensing activities, organizations do not necessarily need to aim for making profits by themselves. Instead, they are required to play certain roles in the entire society. Whether the industrial liaison/technology licensing activities are successful or not will be evaluated from the point if they have produced positive effects in the entire local economy – making a model in which the university takes a role in the process of local activation.

The purpose of the existence of IP management systems (organizations), such as a TLO or IP center, and their roles vary according to the universities or local communities. Therefore, it is extremely risky to generalize whether the existence of TLO or IP center is right or wrong. Rather, it seems necessary to establish the IP management system appropriate for each university or community, while considering the administrative side of the university as well as the existence of other organizations. In order to do so, universities themselves need to make efforts once again to understand the outline of IP management, create a vision for themselves and clarify the purpose of IP management. Also, regardless of the divisions or sections, the government should activate discussions which would support the nature of intellectual management and should deal with the various situations that might occur in the future.

5
Technology Transfer in China

Jianyang Yu

5.1
Introduction

In this chapter I discuss technology transfer in China. After an overview of the status and development of technology transfer in China, I will review the intellectual property (IP) protection of technology, which lays the legal foundation for technology transfer in China. I will then discuss the legal and practical framework of technology transfer in China. Since this book is mainly addressed to readers outside of China, I will focus on cross-board technology transfer (i.e. technology import and export) in China. I hope this chapter will serve as a guide for those who are interested in biotechnology import and export in China.

5.2
Overview

In the nearly three decades since China adopted the reform and open-door policy, technology transfer, particularly technology import, has played an important role to reach the stage of China's successful and strong economic development today.

5.2.1
Technology Import

One characteristic of China's technology import has been that it imports both technologies and the equipment enabling the implementation of the technologies. Historically, when China began carrying out the reform and open-door policy in late 1970s, its economy, the industries and the technology capacity were so devastated by the Cultural Revolution that when importing technologies from abroad, it needed the equipment, sometimes a whole assembly line, to enable it to implement the imported technology. This practice remains so even today, although the government has begun encouraging technology import without,

Technology Transfer in Biotechnology. A Global Perspective.
Edited by Prabuddha Ganguli, Rita Khanna, and Ben Prickril

ISBN: 978-3-527-31645-8

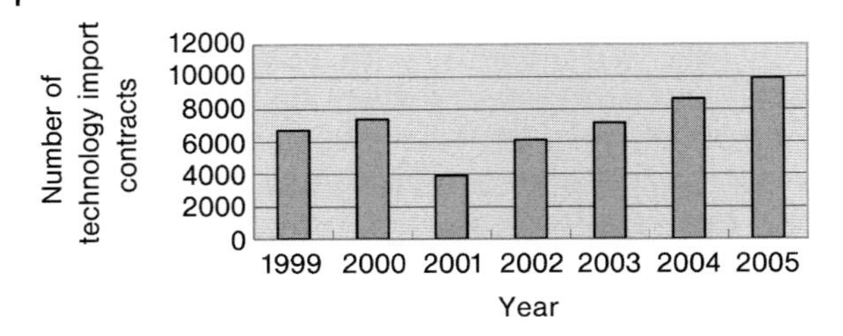

Figure 5.1 Total number of technology import contracts: 1999–2005.

wherever possible, the importation of the related equipment so that precious resources could be invested more in technologies that China needs. Therefore, for almost all the data on technology import released by the government, there is a total amount of contractual money, and then a statement of how much the technology transfer fee is and what is the proportion of the technology transfer fee to the total amount.

According to the statistics from the Ministry of Commerce (MOFCOM), formally the Ministry of Foreign Trade and Economic Cooperation (MOFTEC), China imported almost 50 000 technologies from 1999 to 2005, in which the total amount of contractual money exceeded US$ 100 billion, where technology transfer fees were US$ 62.3 billion (57.6% of the total amount of contractual money).[1] In 2005, the technology import fees were US$ 11.83 billion, which was 62% of the total amount of contractual money of the technology import contracts and is an increase of 31% as compared to the proportion in 1999.[2] MOFCOM has taken it as a positive change of the attitude by enterprises that used to paying more attention to equipment than to technology, and that 'soft' technology imports have gradually become mainstream in China and the quality of imported technology has been noticeably improved.[3] The technology import data for 1999–2005 is given in Figures 5.1 and 5.2 and Table 5.1.[4]

5.2.2
Importing Sources and Industry Dissemination

Taking 2005 as an example year, the European Union was the number one source of technologies imported into China (total: US$ 9.07 billion, including US$ 5 billion from Germany), followed by Japan (US$ 3.85 billion), the United States (US$ 3.4 billion) and South Korea (US$ 0.9 billion). The number one industry

1) News Office of the Ministry of Commerce, *China Imported almost 50,000 Technologies for the Last Seven Years*. www.mofcom.gov.cn, 21 February 2006.
2) Science and Technology Department, MOFCOM, *Historical New Record for Technology Import in 2005*. www.mofcom.gov.cn, 17 January 2006.
3) See Footnote 2.
4) Data released by MOFCOM. www.mofcom.gov.cn.

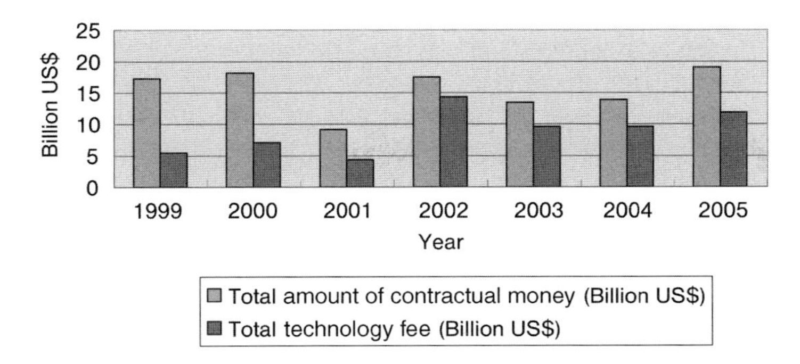

Figure 5.2 Amount of technology import money: 1999–2005.

Table 5.1 Technology import data: 1999–2005.

Year	Total number of technology import contracts	Total amount of contractual money (US$ billion)	Total technology fee (US$ billion)
1999	6678	17.162	5.388
2000	7353	18.176	7.15
2001	3900	9.091	4.395
2002	6072	17.389	14.391
2003	7130	13.451	9.511
2004	8605	13.856	9.625
2005	9902	19.05	11.83

that imported technology in monetary terms was the railway industry (US$ 290 million) in 2005, followed by electronic and telecommunication manufacturing (US$ 210 million), and metallic manufacturing (US$ 196 million).[5)]

5.2.3 Biotechnology

There is no data available relating to biotechnology import for the past years, which means that its monetary amount is not as significant as compared with the top 10 industries listed in the government data. Although a relatively new industry in China, the biotechnology industry is expected to develop significantly in China, particularly in view of the huge market potential. It follows that technology transfer of biotechnology in China will also increase considerably in the near

5) See Footnote 2.

future. As a vice president in one of Shell Group's business unit once put it: 'a distinctive feature of the Chinese market is that it is highly open to advanced technologies, which provides opportunities for Shell to market new products in a high speed'.[6] He may have a point here and it should also apply to biotechnology in China as well.

Also, the national government, the State Council, issued in January 2006 the 'Outline of Long-Term Planning for National Science and Technology Development: 2006–2020', where biotechnology was identified as one of the three strategic science and technology development sections.[7] The State Council states that China needs to put biotechnology as an essential high-technology industry in the future, and needs to strengthen its applications in agriculture, industries, population and healthcare. The outline further lists a number of biotechnology areas as important for future development. It is thus expected that more resources will be poured into biotechnology R&D and its applications in the next 15 years. As a result, there will be more and more biotechnology imports and exports taking place to implement the plan and as a result of the implementation of the plan.

5.2.4
Technology Export

There is no data available on technology export, but it is safe to say that the number of technology export contracts, the total amount of contractual money and the technology fees are much lower than those of the technology imports.

5.2.5
Government Policy

Ever since the adoption of the reform and open-door policy, the Chinese government policy has been encouraging technology transfer, in general and technology import, in particular, as this is an important engine to drive economic development in China. Technology import has been a noticeable factor in formulating the legislations and government policies with respect to foreign trade and foreign investment in China.

Government policies on technology import have also changed over the years. For instance, right after the adoption of the reform and open-door policy, China wanted to import advanced technologies, and the policy has later been changed to importing advanced and appropriate technologies. The government also emphasizes now not only the import of technologies, but the absorption of the imported technologies and technological development based on the imported technologies.

Tax incentives have been provided for the import of advanced technology and of those technologies with favorable contractual terms. Specifically, Chinese tax law

6) International Business Online, *Demands from China Require New Technologies from Multinational Companies*. www.mofcom.gov.cn, 9 March 2006.

7) The State Council, The Outline of Long-Term Planning for National Science and Technology Development: 2006–2020, issued on 9 January 2006.

provides that a foreign technology supplier having no establishment or place in China shall pay an income tax of 20% on technology import income. However, for supply of technical know-how in scientific research, exploitation of energy resources, development of the communications industries, agricultural, forestry and animal husbandry production, and the development of important technologies, the foreign supplier may, upon approval by the competent department for tax affairs under the State Council, be levied at the reduced income tax rate of 10%. Where the technology supplied is advanced or the terms are preferential, exemption from income tax may be allowed.[8] The scope of the reduction or exemption of the income tax is further defined to a number of specific areas, where biotechnology could fall within a number of these areas.[9]

8) Income Tax Law of the People's Republic of China for Enterprises with Foreign Investment and Foreign Enterprises (adopted 9 April 1991, effective 1 July 1991), Article 19.

9) Rules for the Implementation of the Income Tax Law of the People's Republic of China for Enterprises with Foreign Investment and Foreign Enterprises (issued 30 June 1991, effective 1 July 1991), where Article 66 provides that:

The scope of the reduction of or exemption from income tax on royalties provided for in Article 19, paragraph 3, Item (4) of the Tax Law is as follows:

(1) royalties received in providing technical know-how for the development of farming, forestry, animal husbandry and fisheries:
  (a) technology provided to improve soil and grasslands, develop barren, mountainous regions and make full use of natural conditions;
  (b) technology provided for the supplying of new varieties of animals and plants and for the production of pesticides of high effectiveness and low toxicity;
  (c) technology provided such as to advance scientific production management in respect of farming, forestry, fisheries and animal husbandry, to preserve the ecological balance, and to strengthen resistance to natural calamities;
(2) royalties received in providing proprietary technology for scientific institutions, institutions of higher learning and other scientific research units to conduct or cooperate in carrying out scientific research or scientific experimentation;
(3) royalties received in providing proprietary technology for the development of energy resources and expansion of communications and transportation;
(4) royalties received in providing proprietary technology in respect of energy conservation and the prevention and control of environmental pollution;
(5) royalties received in providing the following proprietary technology in respect of the development of important fields of science and technology:
  (a) production technology for major and advanced mechanical and electrical equipment;
  (b) nuclear power technology;
  (c) production technology for large-scale integrated circuits;
  (d) production technology for photoelectric integrated circuits, microwave semi-conductors and microwave integrated circuits, and manufacturing technology for microwave electron tubes;
  (e) manufacturing technology for ultra-high speed computers and microprocessors;
  (f) optical telecommunications technology;
  (g) technology for long-distance, ultra-high voltage direct current power transmission; and
  (h) technology for the liquefaction, gasification and comprehensive utilization of coal.

5.3 Legal Protection of Technology in China

Technology can be protected in China under various legal forms, but the most important forms are patent protection and trade secret protection.

5.3.1 Patent Protection

There are three categories of patent available in China under the Chinese Patent Law:[10] patent for invention, patent for utility model and patent for design. With regard to biotechnologies, it is patent for invention that is relevant. A patent for invention is granted after substantive examination with the term of 20 years of protection.

As for international conventions concerning the protection of patents, China has joined the Paris Convention, the Patent Cooperation Treaty which relates to international patent application procedures and the Budapest Treaty on the International Recognition of the Deposit of Microorganisms for the Purposes of Patent Procedure. China has also joined the World Trade Organization and therefore is a party to the Trade-Related Aspects of Intellectual Property (TRIPS) Agreement.

5.3.1.1 Patent Filing and Prosecution

With the strong economic development in China over recent years and the huge domestic market in sight, patent applications filed by both Chinese and foreign applicants have increased steadily. In 2005, China received 450 000 patent applications and became the number one country in the world in terms of patent applications received,[11] and filed 2452 patent applications abroad and therefore was the number 10 country in the world in this category.[12] Data on patent applications filed in China for 1995–2005 are given in Figures 5.3 and 5.4 and Tables 5.2 and 5.3.[13]

5.3.1.2 Patent Enforcement

A patent holder has two alternatives against patent infringement: either to request an Administrative Authority for Patent Affairs (AAPA) to handle a patent infringement dispute or to bring a civil lawsuit against patent infringement directly with a court.

10) Patent Law of the People's Republic of China (adopted 12 March 1984, effective 1 April 1985, amended 4 September 1992, further amended 25 August 2000), Article 2.

11) China News Agency, *With the Increase of Patent Applications, China Becomes Top Country in the World*, 24 March 2006.

12) China News Agency, *The Number of Patent Applications Files Abroad Increased, Ranking as the No. 10 in the World*, 6 February 2006.

13) Data of patent applications filed from 1999 to 2005 is based on database of the State Intellectual Property Office at http://www.sipo.gov.cn. Data of patent applications filed from 1995 to 1998 is based on the database of China Intellectual Property Net at http://www.cnipr.com.

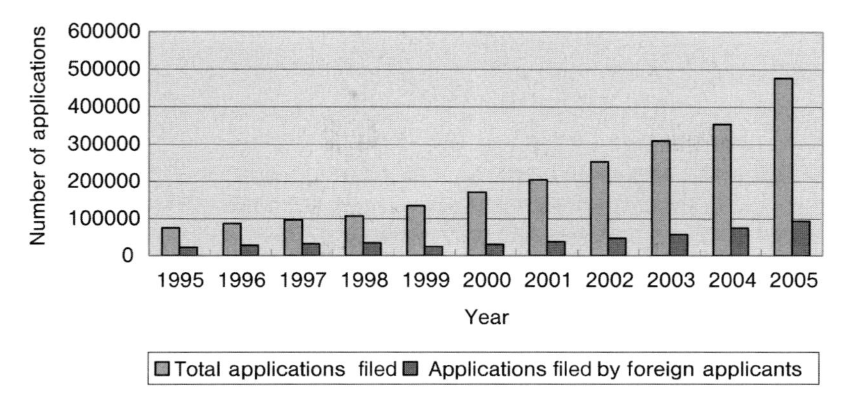

Figure 5.3 Number of patent applications filed in china for 1995–2005.

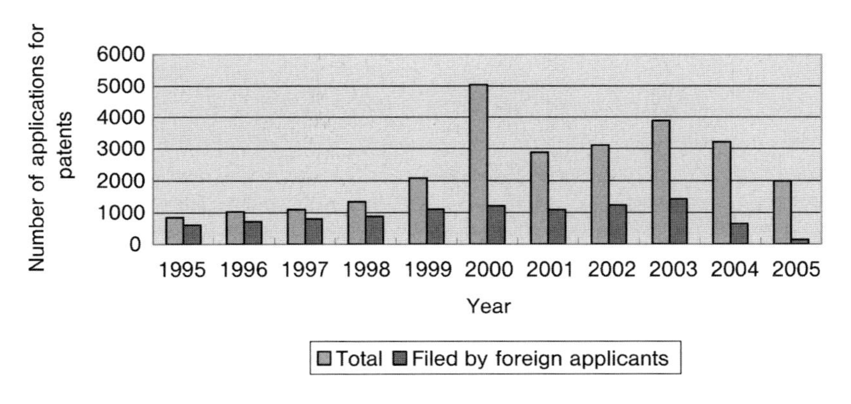

Figure 5.4 Number of applications for patents on biotech for 1995–2005.

The AAPAs have been established at the provincial level and some special cities. The judiciary system in China consists of courts at four levels: the Supreme Court, the higher courts at the provincial level, the intermediate courts at the city level and the local courts at the county level. A Supreme Court Circular grants designated intermediate courts jurisdiction over patent infringement cases as the court of first instance. These designated intermediate courts are those located in the capital of a province or autonomous region, or a municipality directly under the central government, or those specifically designated. The court decision can be appealed to a high court, the decision of which is final.

A number of courts have set up IP divisions to adjudicate, among others, patent infringement cases (and technology contract disputes as well). Where there is no IP division, the economic division of a court will hear patent infringement cases. Usually a panel of three judges will adjudicate a case. A non-judge may be invited by the court to sit on the panel. There is no discovery and no jury trial in China, and court decisions do not enjoy *stare decisis* status.

Table 5.2 Number of patent applications filed for 1995–2005.

Year	Total applications filed	Applications filed by foreign applicants
1995	74549	22448
1996	86550	27481
1997	96675	32122
1998	106850	34548
1999	134239	24269
2000	170682	30343
2001	203573	37800
2002	252631	47087
2003	308487	57249
2004	353807	74864
2005	476264	93107

Table 5.3 Number of applications for patents on biotech for 1995–2005

Year	Total applications filed	Applications filed by foreign applicants
1995	840	602
1996	1022	715
1997	1096	798
1998	1342	877
1999	2086	1104
2000	5037	1217
2001	2876	1094
2002	3099	1236
2003	3881	1423
2004	3199	650
2005	1992	146

Under the Chinese Patent Law, a party who commits any one of the following acts, for production or business purposes without the authorization of a patentee, is liable for patent infringement:

- Making, using, offering for sale of, selling or importing a patented product.
- Using a patented process.
- Using, offering for sale, selling or importing the product directly obtained by a patented process.

Thus, parallel importation of a patented product or a product directly obtained from a patented process constitutes patent infringement in China.

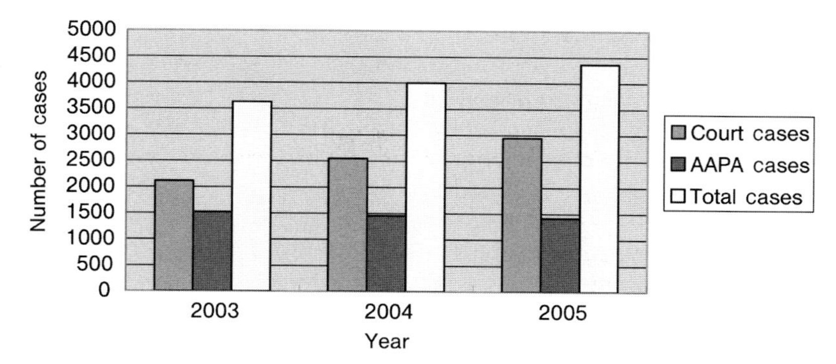

Figure 5.5 Patent dispute cases: 2003–2005.

The legal remedies available to a patent holder include:

- Pre-litigation injunction – a sort of preliminary injunction.
- Permanent injunction.
- Compensation for damages, calculated on one of the following bases:
  - the patentee's actual economic loss caused by the infringement
  - the infringer's total profits derived from the infringement
  - one to three times of patent license royalty
  - 'statutory damages' between about RMB5000–300 000 generally (about US$ 600–37 000), but no more than RMB500 000 (about US$ 62 000).
- Elimination of infringement effect.
- Reasonable expenditure to investigate and stop infringement.
- Court cost paid by loser, but no or very little attorney's fees.

5.3.1.3 **Patent Enforcement Data**

Contrary to tradition and therefore conventional wisdom, Chinese patent holders proved to be determined to enforce their patents by legal means. Data on court and AAPA patent cases for the time period of 2003–2005 is given below and in Figure 5.5:[14)]

- Court cases (2003: 2109 cases; 2004: 2549 cases; 2005: 2947 cases).
- AAPA cases (2003: 1517 cases; 2004: 1455 cases; 2005: 1419 cases, including ownership cases).
- Total: (2003: 3626 cases; 2004: 4004 cases; 2005: 4366 cases).

The above cases are not limited to patent infringement cases and also include other patent dispute cases such as patent ownership disputes. This data shows that China is perhaps the number one country in the world in terms of the

14) All data is from the China Intellectual Property Office (SIPO) at www.sipo.gov.cn, except the data of AAPA cases in 2005 is from http://scitech.people.com.cn/GB/53753/4314743.html.

number of patent disputes cases filed with legal authorities. Among all these patent cases, based on the author's information, the patentee overall win rate in patent infringement actions is about 60%.

There is no data on the number of patent infringement cases involving foreign parties. However, there is data showing that in 2004, courts in the first instance concluded 8332 IP cases where 152 cases involved foreign parties, which is only 1.8% of all the IP cases.[15] The author's assessment is that the percentage concerning patent infringement cases would be around 1.8%, if not less.

Again, there is no data on the win rate of foreign companies, but the author's own experience and observations tend to show that in majority of patent infringement litigation cases, foreign companies have won either as plaintiffs or as defendants.

5.3.1.4 Strategic Considerations

There have been so many cases where foreign companies found, to their regret, that they had failed to file patent applications in China when there were business opportunities presented before them, including transfer of technology and/or when they needed to protect their technology in China that may have worldwide implications. Hence, if technology is to be disclosed by filing patent applications anywhere in the world, and if a foreign company wishes to make use of its technology in China either by technology transfer, arrangements by capital contribution to joint ventures or wholly foreign-owned enterprises, to explore the market directly, or to protect its world market, it is strongly advised that it should file patent applications promptly in China to be in a position to protect its technology and to explore commercial opportunities. That is what an increasing number of multinational companies have been doing – upgrading China to their first tier countries for filing patent applications around the world.

5.3.2 Trade Secret Protection

Trade secrets can be protected in the China under the Contract Law of China ('Contract Law'), Anti-Unfair Competition Law and Criminal Law.[16]

5.3.2.1 Contractual Protection

A trade secret holder may protect its trade secrets by contract. The Contract Law is applicable to transnational contracts where Chinese law is chosen as the governing law. Where appropriate provisions have been included in a contract, the trade secret holder has a remedy for breach of contract.

15) See Footnote 14.

16) Contract Law of the People's Republic of China (adopted 15 March 1999, effective 1 October 1999); Anti-Unfair Competition Law of the People's Republic of China (adopted 2 September 1993, effective 1 December 1993); Criminal Law of the People's Republic of China (adopted 1 July 1979, amended 14 March 1997, effective 1 October 1997).

5.3.2.2 **Protection under the Anti-Unfair Competition Law**

The Anti-Unfair Competition Law provides civil and administrative remedies against infringement of trade secrets. The law defines 'trade secrets,' identifies the infringing acts and sets out remedies and penalties for infringement.

The Anti-Unfair Competition Law defines trade secrets as: '[T]echnological information and business information, that is not known to the public, derives economic value for the holder, is of practical applicability, and has been subject to steps by the holder to maintain its secrecy.'

The Anti-Unfair Competition Law provides the following categories of infringing acts:

- Acquiring a trade secret of another by theft, bribery, coercion or other improper means.
- Disclosing, using or permitting others to use a trade secret acquired by means specified above.
- Disclosing, using or permitting others to use a trade secret of another in breach of an agreement or a duty to maintain secrecy.

In addition to the primary acts of infringement of trade secrets set out above, the Anti-Unfair Competition Law also provides that a third party who knows or should have known that the above acts were illegal, will also be liable for acquiring, using or disclosing those trade secrets.

Where the infringement of trade secrets occurs, the anti-unfair competition Law provides that a trade secret holder has an option either to seek civil remedies by filing a lawsuit with a court, or to request administrative penalties by filing a complaint with an administrative authority for industry and commerce above county level.

Civil remedies include an injunction and awarding compensation for damages. Damages are calculated on the basis of the trade secret holder's actual losses caused by the infringement or, if the losses are difficult to calculate, the amount of damages awarded should be based on the infringer's profit derived from the misappropriation.

As for administrative remedies, there are two sanctions: an injunction and imposing of a fine of RMB10 000–200 000 (US$ 1200–25 000). However, the administrative authority cannot order the payment of damages since it is an administrative action. Judicial review is available for an administrative authority's decision.

5.3.2.3 **Criminal Sanctions**

Criminal law provides criminal sanctions against misappropriation of trade secrets. The maximum penalty is 7 years in prison. The author's information is that approximately 100 criminals were convicted and went to jail each year for the last 5 years or so.

5.3.2.4 **Strategic Considerations**

Where technology transfer relates to licensing-out of trade secrets, the licensor needs to make sure that a protective system should be in place from the very beginning to safeguard their trade secrets. Protective measures should include, for instance, the signing of confidential agreements and restrictive covenants, record-keeping programs, physical security, and education of the licensee's employees.

5.3.3 Other Forms of Protection

There are other legal protections available for technology, where appropriate, which include:[17]

- Copyright Law.
- Regulations on the Protection of Compute Software.
- Regulations on the Protection of Layout-Designs of Integrated Circuits.
- Regulations on the Protection of New Plant Varieties.
- Regulations on Customs Protection of Intellectual Property.

5.4 Technology Transfer in China

In this section I will discuss the framework of technology transfer, focusing on technology import and export for the purpose of this book.

5.4.1 Technology Import and Export

The legal framework of technology import and export in China is mainly defined by the Contract Law, the Supreme Court Interpretations on Several Questions with Respect to the Application of Law when Adjudicating Technology Contract Dispute Cases (hereinafter 'the Supreme Court Interpretations')[18] and the Regulations on the Administration of Technology Import and Export Contracts (hereinafter 'the Import and Export Regulations').[19]

17) Copyright Law of the People's Republic of China (adopted 7 September 1990, effective 1 June 1991, amended 27 October 2001); Regulations on the Protection of Compute Software of the People's Republic of China (issued 20 December 2001, effective 1 January 2002); Regulations on the Protection of Layout-Designs of Integrated Circuits of the People's Republic of China (issued 2 April 2001, effective 1 October 2001); Regulations on the Protection of New Plant Varieties of the People's Republic of China (issued 20 March 1997, effective 1 October 1997); Regulations on Customs Protection of Intellectual Property of the People's Republic of China (issued 5 July 1995, effective 1 October 1995).

18) Interpretations on Several Questions with Respect to the Application of Law when Adjudicating Technology Contract Dispute Cases by the Supreme Court of the People's Republic of China (issued 16 December 2004, effective 1 January 2005).

19) Regulations on the Administration of Technology Import and Export Contracts of the People's Republic of China (issued 10 December 2001, effective 1 January 2002).

5.4.1.1 **Contract Law**

Although a full chapter in the Contract Law is dedicated to technology contracts, there are still many questions unanswered by the law. Therefore, the Supreme Court, which has the legal power under Chinese law to interpreter laws in their applications, issued the Supreme Court Interpretations in 2004 to clarify issues.

5.4.1.2 **Import and Export Regulations**

As for technology import and export, based on the authorization by the Foreign Trade Law,[20] the State Council issued the administrative regulations, the Import and Export Regulations, which became effective on 1 January 2002 and replaced the old Regulations on Administration of Technology Introduction Contracts.[21]

It is important to note that as far as technology import and export contracts are concerned, if there is any discrepancy or conflict between the Import and Export Regulations and the Contract Law/the Supreme Court Interpretations, the Import and Export Regulations will be applied because the Contract Law specifically provides that where the laws and regulations stipulate otherwise on the technology import and export contracts, such provisions shall be followed.[22] Therefore, although the Import and Export Regulations is of a lower level of legislation than the Contract Law, it will prevail if there is any discrepancy or conflict with respect to technology import and export contracts.

The Import and Export Regulations apply to all technology import and export contracts, and provide that:

> The technology import and export as referred to in these Regulations means acts of transferring technology from outside the territory of the People's Republic of China into the territory of the People's Republic of China or visa versa by way of trade, investment, or economic and technical cooperation.
>
> The acts mentioned in the preceding paragraph include assignment of the patent right, assignment of the patent application right, licensing of patent, assignment of technical know-how, technical services and transfer of technology by other means.

It should be noted that although not specifically identified, licensing of technical know-how is also covered under the definition. This provision shows that the Import and Export Regulations govern not only straightforward technology import and export contracts, but also other forms of technology import and export, such as using imported technology as capital contribution to an equity joint venture, using imported technology as investment in a wholly foreign-owned

20) Foreign Trade Law of the People's Republic of China (adopted 12 May 1994, effective 1 July 1994).

21) Regulations on Administration of Technology Introduction Contracts of the People's Republic of China (issued and effective 24 May 1985, abolished 1 January 2002).

22) Contract Law, see Footnote 16, Article 355.

enterprise, and using imported technology in a technical service, a joint research or a development project.

5.4.2
Technology Import and Export: Government Regulations

Technology import and export is governed by government regulations, particularly the Import and Export Regulations. MOFCOM is in charge of the administration of technology import and export nationwide, and the commerce department of governments at provincial or equivalent level are responsible for the administration of technology import and export in their respective regions.

5.4.2.1 Three Categories

The Import and Export Regulations govern all import and export technology contracts, but not all such contracts are required to be examined and approved by government authorities. The Import and Export Regulations differentiate among three categories of technology in relation to import or export:

1. Technology that is prohibited to import or export.
2. Technology that is restricted to import or export.
3. Technology that is free to import and export.

To implement the Import and Export Regulations, MOFTEC has published a number of rules and lists with other government agencies with respect to technologies that are prohibited or restricted for import or export, which are expected to be updated from time to time.[23]

5.4.2.2 Technology Prohibited to Import or Export

The technologies prohibited to import or export are identified in the above lists.

5.4.2.3 Technology Restricted to Import or Export

For technology that is restricted to import, the Chinese recipient needs to file an application for technology import with the government authorities. The government authority will examine the application and decide whether or not to grant the application and issue a Certificate of Intent to Import Technology within 30

23) Rules on Administration of Technologies that are Prohibited or Restricted to Import by the Ministry of Foreign Trade and Economic Cooperation and the State Commission of Economy and Trade (issued 30 December 2001, effective 1 January 2002); List of Technologies that are Prohibited or Restricted for Import issued by the Ministry of Foreign Trade and Economic Cooperation and the State Commission of Economy and Trade (issued on 12 December 2001, effective 1 January 2002); Rules on Administration of Technologies that are Prohibited or Restricted to Export by the Ministry of Foreign Trade and Economic Cooperation and the Ministry of Science and Technology (issued 30 December 2001, effective 1 January 2002); List of Technologies that are Prohibited or Restricted for Export issued by the Ministry of Foreign Trade and Economic Cooperation and the Ministry of Science and Technology (issued 12 December 2001, effective 1 January 2002).

working days. The Chinese recipient can then sign a technology import contract with the foreign technology supplier. The Chinese recipient should submit to the government authorities afterwards with the technology import contract and other document to apply for the Technology Import License, and the government authorities will examine the authenticity of the contract and decide whether to issue the license within 10 working days. Alternatively, the Chinese recipient may choose to apply for the Technology Import License directly with the signed technology import contract, and the government authorities will examine application as well as the authenticity of the contract and decide whether to issue the license within 40 working days. In either case, the technology import contract becomes effective on the day the Certificate is issued.

For technology that is restricted to export, the Chinese supplier is required to file an application for technology export with government authorities. The government authorities will examine the application and decide whether or not to grant the application and issue a Certificate of Intent to Export Technology within 30 working days. Only after obtaining this Certificate can the Chinese recipient begin substantial negotiations with a foreign party and sign a technology export contract. Then the Chinese recipient should apply for the Technology Export License with the government authorities with the following documents:

1. Certificate of Intent to Export Technology.
2. A copy of the technology export contract.
3. An export list of technology documents.
4. Documents proving the legal status of the parties to the contract.

The government authorities will examine the authenticity of the contract and decide whether to issue the Technology Export License within 15 working days. The technology export contract becomes effective on the day the Certificate is issued.

5.4.2.4 **Technology Free to Import or Export**

Any technology that is not listed in the prohibited or restricted list for import or export is free to import or export. For technology free to import or export, the Import and Export Regulations do not require government examination and approval, and only request registration of the contract. The contract becomes effective according to law and the registration of the contract with the government authority is not a prerequisite to make the contract legally effective. When registering such a contract with government authority, the Chinese recipient or supplier should submit to the government authority the following documents:

1. An application for the registration of technology import or export contract.
2. A copy of the technology import or export contract.
3. Documents proving the legal status of the parties to the contract.

The government authority shall register the contract within three working days, and issue the Registration Certificate of Technology Import or Export Contract.

5.4.2.5 License and Registration Certificate

The Technology Import/Export License or the Registration Certificate of Technology Import/Export Contract is an important document. The Chinese technology recipient or supplier needs to present it before government agencies and other institutions, such as the local tax bureau, foreign currency control bureau, bank and customs when handling withholding tax, converting foreign currency for payment of royalties, remittance of royalty to foreign licensor and possible import or export of related equipment.

5.4.2.6 Technology as Investment

Where a foreign party is to use imported technology as investment to set up a foreign-invested enterprise, the procedure for the establishment of the foreign-invested enterprise will be followed with respect to the examination or approval of the import of the technology.[24)]

5.4.2.7 Legal Liabilities and Judicial Review

The Import and Export Regulations provide a number of legal liabilities for violation of the regulations, particularly concerning technologies that are prohibited or restricted for import or export. Also, any decision by the government agency authorized under the Import and Export Regulations can be appealed for administrative reconsideration, or can be appealed to court directly.

5.4.2.8 Import or Export of Dual-Use Technologies

There are a number of government regulations and rules governing the import or export of dual-use technologies with civilian and military applications. These regulations and rules are independent of the Technology Import and Export Regulations and are applied separately.[25)] Specifically, there are regulations on biological products, related equipment and technologies, where a list is attached specifying

24) The Import and Export Regulations, see Footnote 19, Article 22.

25) Regulations of the People's Republic of China on Administration of Chemicals Subjected to Supervision and Control (issued and effective 27 December 1995); Regulations of the People's Republic of China on the Administration of the Nuclear Export (issued and effective 10 September 1997); Regulations of the People's Republic of China on the Administration of the Export of Dual-Use Nuclear Facilities and Related Technologies (issued and effective 10 June 1998); Regulation of the People's Republic of China on Controlling the Export of Guided Missiles and Related Items and Technologies (issued and effective 22 August 2002); Measures for Controlling the Export of Relevant Chemical Products, Affiliated Equipments and Technologies by the Ministry of Foreign Trade and Economic Cooperation, the State Economic and Trade Committee and the Customs General Administration (issued 18 October 2002, effective 19 November 2002); Regulation of the People's Republic of China on the Administration of Precursor Chemicals (issued 26 August 2005, effective 1 November 2005); Measures on the Administration of License for the Export and Import of Dual-Use Products and Technology by the Ministry of Commerce and the Customs General Administration (issued 21 December 2005, effective 1 January 2006).

a number of dual-use biological products, technologies and equipment that fall within the scope of the regulations.[26)]

5.4.3
Technology Import and Export: Important Issues

The Import and Export Regulations apply to all technology import and export, and the Contract Law and the Supreme Court Interpretations may apply to technology import and export if Chinese law is selected as the governing law in a contract. This section will address a number of important provisions under the above legislations.

5.4.3.1 Definition of Technology Contract

A technology contract is defined as 'a contract that the parties conclude for purpose of establishing rights and obligations of the parties regarding technology development, technology transfer, technical consultancy and technical services'.[27)] Technology transfer contracts include contracts on patent transfer, transfer of the right to apply for a patent, transfer of know-how and patent licensing. Contracts on transfer of know-how include contracts on assignment of know-how and licensing of know-how.

5.4.3.2 Forms of Contract

Technology transfer contracts are required to be in written form. The written form means the forms which can show the described contents visibly, such as a written contractual agreement, letters and data-telex including telegram, telex, fax, electronic data exchange and e-mails. Where parties conclude a contract in the form of a letter or data-telex, etc., one party may request to sign a letter of confirmation before the conclusion of the contract and the contract shall be established at the time when the letter of confirmation is signed.

5.4.3.3 Content of Contract

The contents of a technology contract is to be agreed upon by the parties and shall contain the following general clauses:

1. Title of the project.
2. Contents, scope and requirements of the targeted objective.
3. Plan, schedule, time period, place, areas covered and manner of performance.
4. Maintenance of confidentiality of technical information and materials.

26) Regulation of the People's Republic of China on Controlling the Export of Dual-Use Biological Products and Affiliated Equipments and Technologies (issued 14 October 2002, effective 12 December 2002).

27) The Contract Law, see Footnote 16, Article 322.

5. Sharing of liability for risks.
6. Ownership of technological achievements and method of sharing proceeds.
7. Standards and method of inspection and acceptance.
8. Price, remuneration or royalties and method of payment.
9. Damages for breach of contract or method for calculating the amount of compensation for losses.
10. Methods for settlement of disputes.
11. Interpretation of technical terms and expressions.

5.4.3.4 Technology Supplying Party's Obligations

For technology import, there are a number of mandatory obligations imposed on a technology supplying party under the Import and Export Regulations.

The technology supplying party to a technology import contract is required to ensure that it is the legitimate owner of the technology supplied or one who has the right to assign or license the technology. It is also required to ensure that the technology it supplies is complete, accurate, effective and capable of achieving the agreed technical object.

Where the receiving party to a technology import contract is accused of infringement by a third party for using the imported technology, it is required to notify the supplying party immediately, which shall assist the receiving party in removing the impediment. If, however, the receiving party is held liable for infringing the third party's rights by using the imported technology, the supplying party shall bear the liability.

5.4.3.5 Confidentiality and Contract Term

Under the old regulations, unless specifically requested and approved, confidentiality obligations generally should not exceed the term of the technology import contract, which would be 10 years unless specifically requested and approved.[28] Therefore, confidentiality obligations would be only 10 years in general. The Import and Export Regulations have abandoned such a restriction and it is now up to the parties to decide the period of confidentiality in the contract as well as the term of the contract.

5.4.3.6 Restrictive Clauses

The Import and Export Regulations prohibit seven restrictive provisions in technology import contracts under Article 29:

> Technology import contract shall not contain the following restrictive provisions:

28) Regulations on Administration of Technology Introduction Contracts of the People's Republic of China (issued and effective 24 May 1985, abolished 1 January 2002), Article 8; Implementing Rules for the Regulations on the Administration of Technology Introduction Contracts (issued 30 December 1987, effective 20 January 1988, abolished 1 January 2002), Article 13.

(1) requiring the recipient to accept additional conditions which are not necessary for the importation of the technology, including requiring the recipient to purchase unnecessary technology, raw materials, products, equipment or service;
(2) requiring the recipient to pay for or to undertake other obligations for patents which have expired or for technology whose patents have been declared invalid;
(3) restricting the development by the recipient of the imported technology provided by the provider or restricting the use by the recipient of the improvements developed by the recipient;
(4) restricting the acquisition by the recipient of similar or competing technology of the same kind from other sources;
(5) unreasonably restricting the recipient to obtain raw materials, parts and products or equipment from other sources;
(6) unreasonably restricting the quantity, variety or sales price of products produced by the recipient; and/or
(7) unreasonably restricting export channels by the recipient of the products produced by the imported technology.

Thus, any one of the above clauses is invalid, although it may not necessarily make the entire license contract invalid. Interestingly, there are 'six no-nos' under the Supreme Court Interpretations. Article 329 of the Contract Law provides that a technology contract which illegally monopolizes the technology or impedes the technological progress shall be null and void. Article 10 of the Supreme Court Interpretations goes on to list six clauses that are prohibited in technology contracts:

> Technology contracts having the following provisions belong to 'illegally monopolizing the technology or impedes the technology progress' under Article 329 of the Contract Law:

(1) restricting the research and development by one party based on the subject technology or restricting the use of technology by one party improved from the subject technology, or unequal terms for exchange of improved technology between the contracting parties, including requesting one party to provide its improved technology with the other party free of charge, to assign to the other party in non-reciprocal terms, or to make the other party sole licensee or share the IP of the improved technology free of charge;
(2) restricting the acquisition by one party of similar or competing technology from other sources;

(3) preventing one party from fully implementing the subject technology in a reasonable manner according to the market needs, including unreasonably restricting the production of products or service by the recipient using the subject technology with respect to quantity, variety, price, sales channel and export market;
(4) requiring the recipient to accept additional conditions which are not necessary for the implementation of the technology, including requiring the recipient to purchase unnecessary technology, raw materials, products, equipment, services and unnecessary personnel;
(5) unreasonably restricting the recipient to freely obtain raw materials, parts and components, and products or equipment from other channels or sources; and/or
(6) prohibiting the recipient to claim invalidity of the IP rights of the subject technology or attach conditions on such a claim.

There are some overlaps between the 'six no-nos' of the Supreme Court and the 'seven no-nos' of the Import and Export Regulations, but there are still some noticeable differences. For instance, one significant difference is that the Supreme Court Interpretations does not allow a clause that prohibits the recipient to claim invalidity of the IP rights of the subject technology, which is not included in the 'seven no-nos' of the Import and Export Regulations. It should be noted that the 'seven no-nos' of the Import and Export Regulations apply to all technology import contracts, whereas the 'six no-nos' of the Supreme Court will further apply to those technology import contracts where Chinese law is chosen as the governing law.

5.4.3.7 Taxation

There are a number of legislations on taxation relating to technology import and export. As stated before, a foreign technology supplier having no establishment or place in China needs to pay income tax of 20% of technology import income, and the Chinese licensee is required to withhold the tax and pay the tax on behalf of the licensor. Where a foreign technology supplier's country has a bilateral agreement with the China to avoid double tax, the income tax is reduced to 10% of the royalty. Tax incentives for reduction or exemption of the income tax with respect to technical know-how import is discussed above.

5.4.3.8 Governing Law

Parties to a technology import or export contract may choose the law applicable to the settlement of their contract disputes. If the parties have not made a choice, the law of the country to which the contract is most closely connected shall be applied.[29)]

29) Contract Law, see Footnote 16, Article 126.

5.4.3.9 Dispute Resolution

Parties to a technology import or export contract may, according to their arbitration clause or agreement, apply for arbitration to a Chinese arbitration institution or other arbitration institutions. If there is no arbitration agreement between the parties or the arbitration agreement is invalid, a lawsuit may be brought before the court in China.[30]

Where a Chinese court is chosen as the forum to resolve a dispute in a technology import or export contract, the Supreme Court Interpretations provide that the intermediate court should generally be the court of first instance. Where a contract contains contents of both a technology contract and another contract and parties to the contract have disputes in both contents, the case shall be accepted by the court having jurisdiction over the technology contract.[31] If a technology import or export contract is to choose a court in a foreign country to have jurisdiction over any dispute arising from the contract, the parties need to check before signing the contract whether there is any bilateral judicial cooperation agreement between the country and China. If there is no such agreement, enforcement of the court judgment in China, if required, will become a problem, and it would be very difficult, if ever possible, to enforce the court judgment in China absent such a bilateral government agreement.

5.4.3.10 Statute of Limitations

The Contract Law specifically provides that the statute of limitations for action before a court in China or for arbitration before an arbitration institution regarding dispute relating to a contract for technology import or export is 4 years, calculating from the date on which the party knows or should have known the infringement of its rights.

5.5 Conclusions

In conclusion, technology import and export have been thriving in China, and will continue thriving in the foreseeable future. Biotechnology is expected to be one of the most important areas for technology import and export at least for the next 15 years. Legal protection of technology has been strengthened by amendments to legislations and considerable efforts made by central government and enforcement institutions to enforce IP. Government regulations on technology import and export are in place, which are transparent and easy to follow. For foreign technology suppliers, perhaps the first important thing to do is to make sure that their technologies are well covered in China by legal protection of IP.

30) Contract Law, see Footnote 16, Article 128.

31) Supreme Court Interpretations, see Footnote 18, Article 43. A contract dispute generally is within the jurisdiction of a local court and this provision of the Supreme Court grants the jurisdiction to the court of one level higher.

6
Technology Transfer in Latin-America

Claudia Ines Chamas

6.1
Introduction

This work has the objective of analyzing the context of technology transfer in Latin-America with a closer look at the protection and transfer policies developed by Brazil, Argentina and Chile in the field of biotechnology. The subject assumes special importance for the region, which is notably backward technologically. The implementation of the World Trade Organization (WTO)'s Trade-Related Aspects of Intellectual Property Rights (TRIPS) Agreement, specific policies for encouraging academic–corporate partnerships and public–private partnerships as well as a broader comprehension of the complex dimensions of the innovation processes, have led to the reorganization of the innovation models in these countries, which all seek to achieve a much more competitive environment and subsequently greater participation in the global network for leading-edge biotechnological production.

The chapter is organized as follows. Section 2 shows the evolution of the regulations for technology transfer in the context of the relationship between the corporate and academic sectors. Section 3 analyses the support structures for investments in biotechnology in Brazil, Argentina and Chile. Section 4 broaches the marketing policies and practices for biotechnological knowledge. Section 5 concludes with final remarks.

6.2
Academia, Industry and Technology Transfers

Universities, research institutes and industries are areas with different objectives, motivation and cultures. However, the last 30 years have been marked by a significant change in the ways the academic field has inter-related with the business sector and society in general. The advent of the Internet, information technology, and the activities for protecting and economically exploiting intellectual property

Technology Transfer in Biotechnology. A Global Perspective.
Edited by Prabuddha Ganguli, Rita Khanna, and Ben Prickril

ISBN: 978-3-527-31645-8

(IP) have altered the traditional academic dynamic, firmly founded on free access to knowledge. The new dynamic had a profound influence on debates concerning the course of scientific and technological policies throughout the world.

The investments in the protection of knowledge through patents alter the relationship between the scientists and the institutions, corporations and financing agencies. Patent protection for academic knowledge is encouraged with the intent of altering the pace of introducing innovation on the market, accelerating the technology transfer, and offering society new products in a more intensive and continuous manner. On the other hand, affording protection to inventions still in their very early stages may not be appropriate and is under debate despite being a common practice in biotechnology research.

Restrictions in research funding further seem to justify and cause universities to enter the patent business. Ever since the 1980s, the struggle for funds has become increasingly competitive and occurs by means of new selection processes, such as to solve a specific problem or for industry-oriented programs. Venture capital, resources provided by non-profit foundations and university–corporate partnerships seem to compensate for the bitter struggle for government resources. The existent relationships within the academic environment changed over a very short time; the projects under way and research activities soon began to respond to this new dynamic of relationships and financing.[1,2]

Companies are increasingly requiring access to sources of knowledge in order to correct shortcomings or in support of their demands when introducing technological innovations on the market at a constantly faster tempo and, therefore, academic production has become a fundamental element in the more mature systems of innovation. However, other factors come into play: the capability of companies to assimilate and exploit the type of knowledge (often very distinct from something easily placed on the market) produced by the universities and the public research organizations, or the capability of the companies in establishing efficient channels for technology transfers with entities that operate based on the open science model, for example. The increase in academic–industrial cooperation reflects this search for strategic partnerships to make up for this lack of knowledge. Table 6.1 shows some of the motivations for this type of cooperation.

It is interesting to note some evidence of this cooperation, not just in the form of patents, but also as joint publications. Data from the US National Science Foundation show that between 1973 and 1979 the number of scientific articles having joint authorship of scientists from industry and academia increased from 19 to 31%. Further data furnished by Hicks and Hamilton (cited in[3]) lead to the

1) Geuna, A. (2001) The changing rationale for European research funding: are there negative unintended consequences. *Journal of Economic Issues*, **35**, 607–32.

2) Geuna, A. and Nesta, L. (2006) University patenting and its effects on academic research: the emerging European evidence. *Research Policy*, **35**, 790–807.

3) Straus, J. (2000) *Expert Opinion on the Introduction of a Grace Period in the European Patent Law*, European Patent Organization, Munich.

Table 6.1 Incentives for academia–industry cooperation.

Common incentives	Mechanisms for accruing value to the existing scientific and technologic bases of both parts Increasing the amount of patents and publications Improving liaisons between the academic scientists and those from industry Access to dedicated funds for academia–industry cooperation
Specific incentives: industry	Adequate competitive strategy for coping with the high costs and risks of the processes for developing new products and processes Possibility of acquiring access to new markets Reduction of the uncertainties inherent to the processes for generating innovations Correcting faults in the information market Access to libraries and advanced technological resources (equipment, laboratories) Prestige associated with cooperating with highly reputed academic institutions Access to qualified professionals
Specific incentives: academia	Confronting academics with the problems faced by the productive sector Access to knowledge (both tacit and explicit) accumulated by the industrial sector Potential source of jobs for researchers and students alike Circumventing the existing bureaucracy of the government financing agencies Availability of financial resources from the corporate sector for academic research Opportunity for developing new lines of research

same conclusions. Analysis of articles published by US authors between 1981 and 1994, indexed in the Science Citation Index, reveals that the number of articles that included a university or corporation among the authors had more than doubled.

For the universities, the transfer of technology may occur by distinct channels (scientific publications, lectures, licenses for IP rights, professional training, consulting, etc.). Cooperation with other academic institutions and with industries no longer occurs along informal lines, and has become formal, frequent, planned and with relationships governed by contracts that include the regulation of IP rights for intellectual creations that may arise in the scope of cooperative research projects.

Naturally, the environment fostering these partnerships has received new regulatory conditions so as to speed up and make legal the process of adhering to the partnership model as well as creating incentives for the parties, which includes financial incentive for the researchers. The regulatory improvements have established an environment of legal certainty for the risk investors.

The 1980s were marked by profound changes in the policies for appropriation of research results in the United States. The former regime had stimulated retention of ownership by the government or the public domain and, in contrast, there now evolved a new perception that the results of research financed by federal public funds should be used by the private sector in a more intensive manner. Thus, incentive was provided to promote the actual transfer and commercial development of the inventions and other intellectual creations. Biotechnology loomed as a sunrise industry. *Diamond v. Chakrabarty* issued a favorable decision to the inventor in 1980.

The Bayh–Dole Act, proposed by senators Birch Bayh and Robert Dole, allowed universities and small companies to retain ownership of the patents relating to inventions developed with federal government funds. Since then, many universities have established and expanded their technology transfer programs. This trend for increased protection and exploitation of IP is further explained by: the development of biomedical research, which is of great interest to industry; the expansion of the US policy for IP – or the pro-patent era; the long tradition of cooperation between university research and industrial research in the United States; and the characteristics of the US university system (large-scale research, use of diverse sources of financing, both public and private, strong incentives for the researchers to seek out-of-budget resources).[4,5]

Other legislation changes in the United States encouraged the technology transfer process. Among some examples, the more notable are: the Stevenson–Wydler Technological Innovation Act, of 1980, and the Federal Technology Transfer Act, of 1986, that authorized use of the Cooperative R&D Agreements, and the Technology Transfer Commercialization Act, of 2000.

It is interesting to note that both the Bayh–Dole and the Stevenson–Wydler acts established that the royalties from marketing should be shared with the inventor, thus creating a reward system. The legislation allowed the US Government to retain ownership only if it is substantiated that there were no real efforts towards protection. In order to avoid possible abuses of monopoly power, the US Government may issue a license on a non-exclusive basis and retain march-in rights that may be invoked in the public interest. Local industries are also benefited in the case of licensing (35 USC §204, Chapter 18).[6]

4) Granstrand, O. (1999) *The Economics and Management of Intellectual Property: Toward. Intellectual Capitalism*, Edward Elgar, Cheltenham.

5) Mowery, D.C. and Ziedonis, A. (2000) *Numbers, Quality, and Entry: How Has the Bayh–Dole Affected US University Patenting and Licensing?*, Harvard Business School Press, Boston, MA.

6) Christie, A.F., D'Aloisio, S. Gaita, K.L., Howlett, M.J. and Webster, E.M. (2003) *Analysis of the Legal Framework for Patent Ownership in Publicly Funded Research Institutions*, Research Evaluation Programme, Higher Education Group, Commonwealth of Australia, Canberra.

Similar trends are observed in Europe. France approved the Loi sur l' Innovation et la Recherche (Innovation and Research Law) no. 99,587, on 12 July 1999, with the intent of fostering the technology transfer from the public sector to industry and the establishment of technology-based companies. Researchers and university professors from the public sector were allowed to participate in the founding of companies, even as partners or managers, while still retaining their positions as government employees for a period of 6 years. They continue to receive their government salaries during the initial phase of the project.

The United Kingdom adopted a Bayh–Dole style approach for marketing academic inventions. The old National Research and Development Corporation (NRDC), established in 1949, played a special role in the protection of university and public research institute inventions. Later, in the 1980s, the NRDC combined with the National Enterprise Board which gave rise to the British Technology Group. Today, many universities rely on their own structures for protecting and licensing their intellectual creations.

In 1986, Spain promulgated the Ley de Fomento y Coordinacíon General de la Investigacíon y Técnica (Law for the Incentive and Coordination of Research and Technology), which developed the Plan Nacional de Investigacíon y Desarrollo Tecnológico (National Plan for Technological Research and Development). The Oficina de Transferência de Tecnologia (Technology Transfer Bureau) was created within this environment. It organizes and assists the Oficinas de Transferência de Resultados de Investigacíon (Research Results Transfer Bureaus) of the universities, public research foundations and research associations.

In Germany, the Bundesministerium für Bildung und Forshung (BMBF); (Government Ministry of Teaching and Research) launched the BMBF-Patentinitiative in 1996, with the intent of increasing the amount of academic patents and providing incentive for marketing these. The legislation was also modified so as to allow academic institutions to own the inventions developed by them. Prior legislation gave ownership of the invention to the university professor.

In this sense, it is possible to perceive an internationalization of the norm-setting for IP rights. This is occurring not only in terms of international agreements (the TRIPS Agreement, more specifically), but also through the assimilation of IP policies for specific environments, such as universities.

The majority of universities in the United States and, increasingly, the European and Asiatic (South Korea, Taiwan, Japan, etc.) universities, are binding research and joint projects to agreements for the protection and economic exploitation of IP rights, seeking to ensure shared ownership. The transfer of biological material between academic institutions and corporations is also governed by specific agreements. Apart from publications, patents constitute a relevant form of transferring knowledge. It is worth pointing out that patents are temporary rights and have geographical limitations. Thus, anyone interested may access the content of patents from the international patent databases (both free and at a charge). As soon as a patent expires, anyone interested may promote the economic exploitation of the invention described in the patent, since the patent is now in the public domain.

Scientists that until very recently were in complete ignorance of the IP system mechanisms today deal with deadlines and participate in negotiations for the transfer of technology generated within the academic environment. Publication has to be postponed until the invention is sufficiently complete to be described in a patent application. This has led to new forms of rewarding productive scientists through the actual patent.

In this manner, the academic sector is becoming increasingly involved with national innovation systems, interacting with the production sector and becoming an important source of technical expertise and creativity.

6.3
Support Structures for Investments in Biotechnology

Basically, the countries of Latin-America still present incomplete innovation systems with significant differences for each country. The pace and intensity of technology transfer activities as well as the interaction between industry and academia are compatible with the characteristics and limitations of each particular innovation system. Even faced with frailties in terms of science and innovation, there exists a major effort for the assimilation, preparation and adoption of new policies for encouraging the protection and exploitation of IP within academic environments. As shall be seen later, these policies present similarities with the policies adopted in the United States and Europe.

In Brazil, a long period of import substitution led to the build-up of a solid and diversified industrial capability. The strong participation of large foreign corporations accelerated industrialization through technology transfers. Access to technology on the part of the local companies was basically through the import of machinery and equipment. Notable advances were made in the oil and gas, mining, telecommunications, and aeronautical industries. The import substitution policy and the strongly protectionist approach ended in the 1980s, together with the resurgence and consolidation of democratic institutions and a public finances crisis. The following decade was marked by a process of economic opening and an upgrade of the productive structures, principally aimed at increasing productivity through relatively small investments in the improvement to quality systems. The government program with the broadest scope was the Programa Nacional de Qualidade e Produtividade (National Quality and Productivity Program). The adjustments were mainly intended for the local market.[7,8]

7) Pacheco, C.A. (2003) *Políticas Institucionales en Materia de Propiedad Intelectual y Transferencia de Tecnología. Experiencias Prácticas sobre Mecanismos Institucionales de Vinculación Universidad-Empresa*, OMPI-CEPAL/INN/SAN/03/T2.2b, http://www.wipo.int/meetings/en/doc_details.jsp?doc_id=18361.

8) Guimaraes, J.A. (2004) Medical and biomedical research in Brazil: a comparison of Brazilian and international scientific performance. *Ciência e Saúde Coletiva*, **9**, 303–27.

The changes occurring in the 1990s were, however, insufficient when relating to gains in competitive capacity, through corporate strategies for innovation, technological partnerships and partnerships with universities or research institutes, etc.[7] Furthermore, the assimilation of IP strategies was limited both to the local companies and the universities and research institutes. The volume of venture capital for technological activities was discrete.

Data from 2005 shows that investments in R&D were the equivalent of 1.2% of the gross domestic product, with 60% of this amount being from government funding. Data from 2003 indicates that Brazil had a participation of 1.4% in international scientific production (indexed publications). Medical research produced 7365 articles between 1997 and 2001 (0.9% in this area worldwide), which ranks 23rd in the world and third internally, representing 16.9% of the total articles indexed for the country on the basis of the ISI Standard. The biomedical area showed slightly higher output than the medical area, with 8366 articles for this period (0.9% in this area worldwide), securing Brazil 21st place in the world ranking and second place internally, representing 19.0% of all the country's articles indexed on the basis of the ISI Deluxe.[9] The country also presents advances in the formation of human resources. In 2003, approximately 8000 doctorates were obtained in Brazil. In 2006, this title should be attained by 10 000 individuals.

Since 2000, several measures have been debated and implemented with the aim of encouraging partnerships between the public and private sectors, as well as divulging knowledge about IP and facilitating generation of innovation throughout the productive chain. All governed under the same law, 16 Fundos de Apoio ao Desenvolvimento Científico e Tecnológico (Support Funds for Scientific and Technological Development) were set up for negotiating projects in strategic areas (power, oil and gas, water resources, aerospace, health, agriculture, biotechnology, aeronautics, minerals, telecommunications, transportation, sea and river transportation, the Amazon region, and computer technology). The other two funds are for research infrastructure and for university–corporate interaction.

In March 2004, the Brazilian government instituted the Política Industrial, Tecnológica e de Comércio Exterior (Industrial, Technological and Foreign Trade Policy). Biotechnology was considered the most promising area. This led to the Fórum de Competitividade do Setor de Biotecnologia (Forum for Competitiveness in the Biotechnology Sector).

Brazilian biotechnology is the result of investments that became increasingly important as of the seventies. The Conselho Nacional de Desenvolvimento Científico e Tecnológico (CNPq; National Council of Scientific and Technological Development), which is the government agency for encouraging research, detected serious failings in the existent state of the basic biological sciences within Brazilian academics. There were few productive groups, low multidisciplinary levels, few doctorates in the management of research projects and scant national or

9) Ávila, J. (2004) Algumas considerações sobre os ambientes de inovação nos Estados Unidos e no Brasil. *Comciencia*, http://www.comciencia.br/reportagens/2004/08/10.shtml#_ftn1.

international exchange activity. Other analyses also observed the growing importance of genetic engineering on the world scene. The area called for development; however, there were obstacles both at corporate levels (little interest on the part of companies in internalizing R&D) and at academic levels (with few dedicated biological and biochemical researchers).

The response to these structure problems included the establishment of Centros Integrados de Biotecnologia (Integrated Biotechnology Centers) and technological poles and parks. The CNPq and the Financiadora de Estudos e Projetos (FINEP; Research and Projects Financing) implanted lines of funding with academic objectives: the Programa Integrado de Genética (Integrated Genetic Program), in 1975, the Programa Integrado de Engenharia Genética (Integrated Genetic Engineering Program), in 1978, and the Programa Integrado de Doenças Endêmicas (Integrated Endemic Diseases Program), between 1973 and 1985.[10)]

The outlook of industrial applications first appeared in the 1980s with the Programa Nacional de Biotecnologia (National Biotechnology Program) and with the Subprograma de Biotecnologia (Biotechnology Subprogram), of the Programa de Apoio ao Desenvolvimento Científico e Tecnológico (Scientific and Technological Development Support Program). The development of new products in the health field was expected (such as proteins, vaccines, enzymes with industrial uses, etc.). The projects involved the characterization of antigens and the molecular cloning of parasite genes aimed at the development of vaccines as well as isolating genes and preparing monoclonal antibodies to be used in diagnostic methods. Notwithstanding the intention of providing incentive to the industrial sector, the main beneficiaries of the resources were the universities and research institutes.

The incentive for biotechnology was consolidated at the end of the 1990s, and led to the creation of numerous financing opportunities by all agencies at federal, state and municipal levels. The financing was in the form of regular lines or for special projects, directed at specific subjects. Some of the more noteworthy recent programs are: the Fundo Setorial de Biotecnologia (Sectorial Fund for Biotechnology) and the Fundo Setorial de Saúde (Fund for the Health Sector), under the auspices of the FINEP and the CNPq/Ministério da Ciência e Tecnologia (Ministry of Science and Technology); the Structural Molecular Biology Network (SMOLBnet), the Programa de Pesquisas em Caracterização, Conservação e Uso Sustentável da Biodiversidade do Estado de São Paulo, BIOTA/FAPESP, O Instituto Virtual da Biodiversidade (The Research Program on Characterization, Conservation and Sustainable Use of the Biodiversity of the State of São Paulo, BIOTA/FAPESP, The Virtual Institute of Biodiversity) and the Programa Genoma (Genoma Programm), under the auspices of the Fundação de Amparo à Pesquisa do Estado de São Paulo (Foundation for the Support of Research in the State of São Paulo); several funding calls under the auspices of the Departamento de Ciência e Tecnologia do Ministério da Saúde (Department of Science and Tech-

10) Azevedo, N., Ferreira, L.O. and Kropf, S.P. (2002) scientific research and technological innovation: the brazilian approach to biotechnology. *Dados*, **45**, 139–76.

nology of the Ministry of Health); RioGene and Rede Proteômica (Proteomics Network), under the auspices of the Fundação de Amparo à Pesquisa do Estado do Rio de Janeiro (Foundation for the Support of Research in the State of Rio de Janeiro); amongst others.

Brazilian research in biotechnology has achieved great advances in important areas such as genomics, proteomics, bioinformatics, nanobiotechnology and stem cells. As an example, genome networks in the state of Minas Gerais were able to sequence the genome of the Schistosoma mansoni parasite by processing the information in the Núcleo de Bioinformática (Bioinformatics Nucleus) of the Universidade Federal de Minas Gerais (Federal University of Minas Gerais). Schistosomiasis is endemic in the state with almost 1 million people infected. One of the projects within the scope of the proteomics network of the state of Rio de Janeiro involves the characterization of the regulation of gene expression in pathogenic strains of *Vibrio cholerae*, presently underway at the Universidade Federal do Rio de Janeiro (Federal University of Rio de Janeiro). The use of controlled release systems in pharmaceuticals is one of the subjects of interest to the Rede Nacional de Nanobiotecnologia (National Nanobiotechnology Network), coordinated by the State University of Campinas. In the field of research with stem cells, ongoing clinical tests with 1200 patients afflicted by four different cardiopathies are investigating the effectiveness of implanting this type of cell.

Genomic research in Brazil received great impulse in December of 2000, due to the launch of the Projeto Genoma Brasileiro (Brazilian Genome Project) with the participation of approximately 25 molecular biology laboratories located in all the country's geographical regions and coordinated by the Ministério da Ciência e Tecnologia (Ministry of Science and Technology). Several state networks were subsequently established, which reinforced the system. The results of one of the pilot projects (the sequencing of the *Xylella fastidiosa* bacteria genome) became the cover story of *Nature*. This recognition of scientific merit encouraged research groups throughout Brazil to invest in highly complex research.

Certain public institutions have developed broad skills in managing the diverse disciplines related to biotechnology in health, notably the FIOCRUZ, the Instituto Butantan (Butantan Institute) and several university departments. The Empresa Brasileira de Pesquisa Agropecuária (Brazilian Agricultural Research Company) has shown great articulation in agribusiness, both on the national and international scene.

In Argentina, the first generation of institutions linked to science and technology date back to the fifties, including such organizations as the Comisión Nacional de Energía Atômica (National Commission for Atomic Energy) (1956), Instituto Nacional de Tecnología Agropecuaria (National Institute for Agriculture and Livestock Technology) (1956), Instituto Nacional de Tecnología Industrial (National Institute for Industrial Technology) (1957) and Consejo Nacional de Investigaciones Científicas y Técnicas (National Council for Scientific and Technological Research) (1958). These institutions, in association with the universities, were the mainstay of the system for over four decades. Activities were always supported overall by government inputs, more especially through the formation of human resources, the development of the basic sciences and direct intervention in key sectors (nuclear or aerospace). Much as with the Brazilian system, there was little articulation between investments in research and efforts at modernizing corporate technology.[11] Direct foreign financing and an emphasis on the import of equipment led companies to make efforts to adapt primarily intended at the development of engineering and industrial design capabilities.[12]

In the 1990s, however, the overvalued local currency meant equipment and inputs were freely imported. Institutions emerged for facilitating the process of acquiring and applying knowledge favoring corporate technological modernization, including investments in biotechnology, such as:

1. Unidades de Vinculación Tecnológica (Technological Cooperation Units), attributed with helping companies to develop projects for improving productive and commercial activities. They encourage innovation, technology transfers and technical assistance.
2. Fondo Tecnológico Argentino (Argentine Technological Fund), that finances innovation projects and provides support to: technological development, technological modernization, patenting, technological services for research institutes and companies, qualification, technical assistance and technological consulting, and the incubation of companies, technological parks and poles.
3. Fondo para la Investigación Científica y Tecnológica (Scientific and Technological Research Fund) that supports projects and activities with the purpose of generating new scientific and technological knowledge – both for basic and applied themes – developed by researchers from public and private non-profit institutions.
4. Agencia Nacional de Ciencia y Tecnologia (National Science and Technology Agency), which is an agency of the Ministerio de Educación, Ciencia y Tecnología (Ministry of Education, Science and Technology) dedicated to promoting activities relating to science, technology and productive innovation.

In support of biotechnology, furthermore, the Ministerio de Economía y Producción (Ministry for the Economy and Production), through the Secretaría de Industria y Comercio (Secretariat for Industry and Commerce), Secretaría de la Pequeña y Mediana Empresa (Secretariat for Small and Medium Companies) and the Secretaría de Agricultura, Ganadería, Pesca y Alimentos (Secretariat for Agriculture, Livestock, Fishing and Food) offers mechanisms for the development of productive sectors employing biotechnology. The Instituto de Tecnología Industrial (Industrial Technology Institute) and the Instituto de Tecnología Agropecuária (Agriculture and Livestock Technology Institute) should also be highlighted. The Foro Argentino de Biotecnologia (Argentine Biotechnology Forum),

11) Chudnovsky, D. (1999) Políticas de ciencia y tecnología y el sistema nacional de innovación en la Argentina. *Revista de la CEPAL*, **67**, 153–71.

12) Anlló, G. and Peirano, F. (2005) *Una mirada a los sistemas nacionales de innovación en el Mercosur: análisis y reflexiones a partir de los casos de Argentina y Uruguay*, CEPAL, Buenos Aires.

which is a non-profit organization, represents the greater part of Argentine companies and entities active in the field of biotechnology.

There are over 115 centers, institutes and research groups answering to universities and national institutes, as well as technical groups from private enterprise that perform R&D activities in the field of biotechnology, of which 68 (41 companies and 27 research groups) are dedicated to agriculture and livestock.

There exists today a sound perception of the importance of a joint engagement by the public and private sectors for the success of actions relating to innovation and technology transfers in the field of biotechnology. The government policy outlined in the Plan Nacional de Ciencia y Tecnología (National Plan for Science and Technology) seeks to follow this course. In practice, however, the dynamic of the Argentine innovation system has not yet been substantially altered to the point of allowing an integration that reflects the requirements of a complex innovation system.

Chile repeats a Brazilian and Argentine characteristic – more than 85% of researchers work in universities or research institutes and less than 15% of investments in R&D occur with corporations. The country has a limited amount of local technology producers. Santibáñez[13] explains that industries such as mining, paper and cellulose, winemaking, telecommunications, and salmon fisheries remain technologically updated through the purchase and acquisition of technology available on the international market, whenever possible. This strategy is very limiting since the technology made available on the market may be very far from the leading edge technology of the moment.

The government of Chile instituted its present scientific and technological policy in the 1990s. Its three main objectives are: (i) to strengthen the scientific and technological capability of the universities and technological institutes; (ii) to contribute to the increase in competitiveness of the economical and social sectors of the country; and (iii) to induce R&D within Chilean companies and incite them to greater commitments with innovation. Between the early 1980s and the end of the 1990s the following organizations emerged: Fondo Nacional de Desarrollo Tecnológico y Productivo (National Fund for Technological and Productive Development), Fondo de Fomento al Desarrollo Científico y Tecnológico (Fund for the Fostering of Scientific and Technological Development), Fundación para la Innovación Agraria (Foundation for Agrarian Innovation), Fondo Nacional de Desarrollo Científico y Tecnológico (National Fund for Scientific and Technological Development) and the Fondo de Desarrollo e Innovación (Fund for Development and Innovation). In the scope of the Corporación de Fomento de la Producción (CORFO; Corporation for Fostering Production), which is the Chilean government organ attributed with promoting national productive development, Innova Chile was created in 2004. This new agency answering to CORFO promotes and facilitates innovation to improve the competitiveness of the Chilean economy.

13) Santibáñez, E. (2003) *Sistema Nacional de Innovación y Vinculación Sector Público Privado. Caso de Chile*, Reunión Regional OMPI-CEPAL de Expertos sobre el Sistema Nacional de Innovación: Propiedad Intelectual, Universidad y Empresa, WIPO/ECLAC, Santiago.

One of the main areas handled by Innova Chile is the Área de Difusión y Transferencia Tecnológica (Technological Diffusion and Transfer) in support of the initiatives destined to seek, acquire, adapt and divulge management or production technologies for the corporations.

Innova Chile also contemplates biotechnology. The aim in this area is to increase biotechnological development in the sectors of forestry, agriculture and livestock in order to improve competitiveness, product quality and processes by means of (i) financing biotechnological projects, (ii) forming human resources and (iii) reinforcing the national policy for biotechnology.

Overall, in Latin-America, the innovation systems linked to the biotechnology business show some common frailties: difficulties in attracting private capital, difficulties in establishing appropriation strategies for the entire scope of intangible assets (patents, trademarks, cultivars, etc.), insufficient venture capital, poor liaison between research, innovation and the productive sector, and reduced participation in international research and innovation networks.

6.4 Policies for Marketing Biotechnological Knowledge

The movement for harmonizing IP rights has attained Latin-America, which participated in the negotiation process for the TRIPS Agreement which culminated in its approval in the 1990s. The TRIPS Agreement represents an initiative on the part of the developed countries towards increasing protection for IP. This initiative occurred in the context of international trade expansion and the technological content of these exports, as well as the consolidation of a new directive for global production in which the control of technology gains a differentiated qualitative dimension when compared to the environment in which the Paris Convention was signed in 1883 and its later revisions. The post-TRIPS era also shows a general trend for strengthening the power of the patent owners.[14,15]

The alleged advantages for gains to be derived from increased IP protection, principally for developing countries such as those from Latin-America, were not confirmed. It is interesting to note that when theses countries accepted including the TRIPS Agreement in the Uruguay Round the major gain sought was access to the markets of developed countries for their textile products and clothing, agricultural, and tropical produce as well as other products, but this was not to occur.

However, two gains appear to be unequivocal: retaining the compulsory licensing feature and the possibility of adopting the parallel imports mechanism within the scope of national legislation together with the use of panels in the sphere

14) Chamas, C.I. (2005) *Developing Innovative Capacity in Brazil to Meet Health Needs.* Commission on Intellectual Property, Innovation and Public Health, Geneva.

15) Chamas, C.I., Carvalho, S.P., Salles-Filho, S. and Pinheiro, A.A. (2005) The dynamics of intellectual protection for biotechnology in Brazil, presented at the *5th Triple Helix Conference*, Turin.

of the WTO, within a precise sector for discussing sanctions which therefore no longer take a unilateral character.

Both the restrictions and the advantages tend to vary with the technical and scientific capabilities of the various developing countries, together with their capacity for devising and implementing policies relating to IP and innovation as well as the level of participation of each country in international trade. It is worthy of note that there are very heterogeneous capabilities in education, technology and industry in the countries grouped together under the label 'developing countries'. Morel *et al.*[16] classify the innovative developing countries as a group with substantial capacity for innovation and production. This includes Brazil, India and China, amongst others.

Thus, for example, if the maintenance of compulsory licensing may be understood as a gain for developing countries, this option can be used to better advantage by countries that have the capability of copying a medicine and producing it. This depends on prior investments in the national technical and scientific capability, as well as in the industrial park. It will be very difficult for countries without this capability, previously achieved in a context of fewer restrictions in the access to technology, to attain it under the post-TRIPS conditions of harsher restrictions and therefore more vulnerable from the point of view of technology access.

As of the 1990s, Brazil implemented the renovation of its IP policies. Prior to the present Industrial Property Law of 1996 (Law no. 9279, passed in 1996 and in effect as of May 1997), Brazil had already reformed its legislation concerning the subject, instituting the Industrial Property Code (Law no. 5772, of 21 December 1971). This Code prohibited the patenting of chemical products, food and chemical/pharmaceutical products or processes and did not recognize transgenic microorganisms as patentable subjects. Due to Clause 27 of the TRIPS Agreement, the new Industrial Property Law recognized these fields as patentable matter. However, Brazil has opted for a *sui generis* protection – Plant Variety Protection – for plants.

The Serviço Nacional de Proteção de Cultivares (National Plant Varieties Protection Service), created by Law no. 9456, of 25 April 1997, and subordinate to the Ministério da Agricultura, Pecuária e Abastecimento (Ministry of Agriculture, Livestock and Food Supply) is accountable for the administration of plant variety protection.

With the publication of Provisional Measure no. 2186-16, of 23 August 2001, the legislation relating to genetic assets was altered with respect to the conservation of biological diversity, the integrity of genetic assets and associated traditional knowledge. As of Provisional Measure no. 2186-16 and Decree no. 3945/2001, the

16) Morel, C.M., Acharya, T., Broun, D., Dangi, A., Elias, C., Ganguly, N.K., Gardner, C.A., Gupta, R.K., Haycock, J., Heher, A.D., Hotez, P.J., Kettler, H.E., Keusch, G.T., Krattiger, A.F., Kreutz, F.T., Lall, S., Lee, K., Mahoney, R., Martinez-Palomo, A., Mashelkar, R.A., Matlin, S.A., Mzimba, M., Oehler, J., Ridley, R.G., Senanayake, P., Singer, P. and Yun, M. (2005) Health innovation networks to help developing countries address neglected diseases. *Science*, **309**, 401–4.

access to and dispatch of genetic assets existing in the country is determined by the Council for the Management of Genetic Assets, whereby the benefits are liable for distribution; the exchange and the dissemination of components of genetic assets as well as associated traditional knowledge of the indigenous and other local communities are preserved, provided it is to their benefit and is based on common practice.

In 2004, influenced by certain initiatives of countries such as the United States (the Bayh–Dole Act and others) and France (Loi sur l' Innovation et la Recherche), the Lei da Inovação (Innovation Law) (Law no. 10.973, of 2 December 2004), was approved. This should have great impact on biotechnology. An increase in the partnerships between companies, universities and scientific/technological institutes is expected. The possibility of attracting university researchers to start companies dedicated to innovation is another strong point. It serves as a stimulus to the creation of technology-based companies, capable of marketing the results of research undertaken in universities and research institutes. The participation of these researchers in the management or administration of private companies is now allowed, underlining the entrepreneurial potential of these professionals. It also allows sharing the space and infrastructure of public research with private companies. The law advances the elimination of various bureaucratic hindrances, such as the requirement of a bidding process for the licensing of patents when these belong to a public agency.

The Innovation Law calls for the establishment of Technological Innovation Offices at universities and research centers. This novel and powerful incentive is expected to encourage the protection and commercialization of academic inventions, fostering economic dynamism and new job opportunities. Since the mid-1990s, many universities and research institutes have organized themselves to offer IP services. The inventions generated in these organizations belong to the employer in accordance with the Industrial Property Law.

Although the number of patents constitutes a rather incomplete indicator for gauging technological innovation, it remains a useful factor for following trends. The number of patents for Brazilian inventions reflects poor overall capability for innovation and that the players in the innovation process are unprepared to handle the multiple aspects of IP. Furthermore, the protection process, principally abroad, remains rather expensive for local organizations. In the field of IP, a mere 192 patents were granted to Brazilian inventors by the US Patent and Trademark Office (PTO) in 2004. The numbers between 2000 and 2003 are equally modest: 136 patents (2000), 165 patents (2001), 150 patents (2002) and 209 patents (2003). The numbers relating to Patent Cooperation Treaty (PCT) patent applications are not encouraging either (see Table 6.2): 178 applications (2000), 173 applications (2001), 201 applications (2002), 219 applications (2003) and 281 applications (2004).

Brazil offers two outstanding examples of IP policy applied to biotechnology: the role of EMBRAPA (the Agricultural Research Institute of the Ministry of Agriculture) in the Brazilian plant seed market and the role of FIOCRUZ (a Ministry of Health institute that works with research, education, technological develop-

Table 6.2 PTO patents and PCT applications: Argentina, Brazil and Chile

Country		Year				
		2000	2001	2002	2003	2004
Argentina	PCT applications	9	9	9	10	15
	PTO patents	73	62	64	76	66
Brazil	PCT applications	178	173	201	219	281
	PTO patents	136	165	150	209	192
Chile	PCT applications	1	3	6	9	6
	PTO patents	18	20	18	16	21

Research based on PTO patents listed by inventor country.

ment and production in the field of the human health) in the 'drug cocktail' for the AIDS Program of the Brazilian government. In the first case, supported by an IP policy in the area of plant varieties, EMBRAPA was able to assemble partners, both public and private, who worked on the development of new plant varieties, allowing the country to keep the majority of national plant varieties after the promulgation of the Plant Variety Protection Law in 1997, pursuant to TRIPS requirements.

FIOCRUZ, through Farmanguinhos, its drugs production unit, provided the Ministério da Saúde (Ministry of Health) with the cost structure of the drugs that constitute the 'drug cocktail' used in the AIDS Program, identifying the necessary technology for its production.[14,15)]

In both FIOCRUZ and EMBRAPA a new standard of research organization is being implemented. The search for partnerships and the sharing of proprietary results as well as the search for complementary competencies, which would be impossible to find in a single research institution or national economic agent, is a main factor. The rationale underlying the role of public research may be centered in the relevant markets, while retaining focus on the mandate and rationale for the generation of technical and scientific knowledge.[17)]

Experience with technology transfer processes for immunobiologicals and biopharmaceuticals in the field of health must be stressed. In Brazil, universal access (public and gratuitous) to health services is a constitutional right and this is implemented by means of decentralized actions, either at state or municipal levels, with the technical and financial cooperation of the federal government. In order to render the programs for the distribution of medicines and vaccines to the needy population financially viable, the Ministry of Health seeks international

17) Salles-Filho, S. (ed.) (2000) *Ciência, tecnologia e inovação: a reorganização da pesquisa pública no Brasil*, Komedi/Capes, Campinas/Brasília.

partnerships to promote technology transfers enabling the manufacture of all products locally.

The production of vaccines dates back to the beginning of the twentieth century, starting with the foundation of the Butantan Institute and the Instituto Soroterápico Federal (Federal Serum-Therapeutic Institute) (now FIOCRUZ). Since 1973, the government develops vaccination strategies (chiefly through national vaccination day campaigns) that reach all Brazilian municipalities. Up until the end of the seventies, the supply of vaccines issued from the private sector or imports. With the closure of Sintex do Brasil in the 1980s, the Ministry of Health decided to strengthen this area through a program that would stimulate local capability, since not all could be imported (certain antiophidic serums, for example, due to the specificity of the venoms). Currently, the production of vaccines in Brazil is concentrated in the public administration. Since 1986, more than US$ 150 million have been invested in the modernization of the installations and equipment of the public laboratories producing serums and vaccines, within the scope of the Programa de Auto-Suficiência Nacional em Imunobiológicos (PASNI; National Program for Self-Sufficiency in Immunobiologicals) of the Ministry of Health. PASNI was created in 1985 with the specific purpose of strengthening the productive industrial park and establishing a policy of national production.

Brazil is one of the biggest markets in the world for vaccines and relies on a far-reaching immunization program. It has attained self-sufficiency in the production of antiophidic serums, antivenins, antitoxics for therapeutic use and eight vaccines: Bacille Calmette-Guérin; poliomyelitis, recombinant hepatitis B; diphtheria; tetanus; whooping cough [diphtheria, tetanus and pertussis (DTP)]; yellow fever; Haemophilus influenzae type b (Hib), dispensed jointly with DTP; and influenza (for the elderly).[18] Two local manufacturers – Unidade de Produção de Imunobiológicos da Fiocruz (Immunobiological Production Unit) of FIOCRUZ and Butantan Institute – together concentrate 89% of all sales to the Ministry of Health. The other two manufacturers are Instituto de Tecnologia do Paraná (Paraná Technology Institute) and the Fundação Ataulpho de Paiva (Ataulpho de Paiva Foundation). Sanitation control in the field of vaccines is the responsibility of the Agência Nacional de Vigilbncia Sanitária (National Health Surveillance Agency).

A more recent approach emphasizes the formation of strategic alliances for the production of vaccines in Brazil. This is the case of the tetravalent DTP + Hib vaccine, whereby the DTP is produced by the Butantan Institute and is dispensed jointly with the Hib vaccine made at BioManguinhos.

The Hib vaccine was obtained through cooperation between the Manguinhos Immunobiological Unit (BioManguinhos) of the Oswaldo Cruz Foundation (FIOCRUZ) and GlaxoSmithKline. The national production of this formerly imported vaccine now means that 100% of the Ministry of Health demand intended

18) Homma, A., Martins, R.M., Jessouroum, E. and Oliva, O. Desenvolvimento tecnológico: elo deficiente na inovação tecnológica de vacinas no Brasil. *História, Ciências, Saúde – Manguinhos*, **10** (Suppl 2), 671–96.

for the basic vaccination calendar will be met and represents a direct savings of US$ 3.7 million per annum for Brazil. The import of vaccines rose from the level of US$ 70 million (1997/1998) to 125 million (1999/2001).[19]

Starting in 2008, BioManguinhos will produce and supply the Sistema Único de Saúde (Brazilian Single Health System) with two medicines currently imported from Cuba. One is recombinant human α-erythropoietin (EPO), which is used in the therapy for anemia associated with renal insufficiency, AIDS and chemotherapy. The other is recombinant human α-interferon (IFN), which is adopted in the treatment of viral hepatitis. Local production will become possible due to a technology transfer process with Cuba. An annual production of 7.5 million vials of EPO and IFN is expected.

In Argentina, the revision of the IP legislation included the approval of Law no. 24.481 for invention patents and utility models, and Decree 260/96, which established a regulatory mark for the patenting of biotechnological processes and products. Law no. 24.481 was later modified by Law no. 24.572. In accordance with Clause 10 of the Argentine patent law, the owner of the patent is the employer. Argentina also possesses a law for the protection of plant seeds and phytogenetic creations.

In the field of IP, 66 patents were granted to Brazilian inventors by the PTO in 2004. The numbers between 2000 and 2003 were: 73 patents (2000), 62 patents (2001), 64 patents (2002) and 76 patents (2003). The numbers relating to PCT patent applications do not represent intensive filing activities (see Table 6.2): nine applications (2000), nine applications (2001), nine applications (2002), 10 applications (2003) and 15 applications (2004).

Organizations linked to the area of science and technology seek to establish services for IP. This is the case of the Consejo Nacional de Investigaciones Científicas y Técnicas (National Council for Scientific and Technological Research) which is the main organ dedicated to the promotion of science and technology in Argentina, which established the Dirección de Vinculación Científico-Tecnológico (Scientific-Technological Cooperation Administration). This organ has the purpose of advising researchers on the protection of the results of their research.

In Buenos Aires, the Dirección de Convênios y Transferência de Tecnologia (Technology Transfers and Agreements Administration) which was the result of Resolution no. 647, of 1987, provides management for the transfer of research results.

The protection of plant varieties is covered by Law no. 20.247, of 1973, for plant seeds and phytogenetic creations and by Decree no. 2183/91 which serves to regulate it. Law no. 24.376, of 1994 approved the UPOV Agreement (International Union for the Protection of New Varieties of Plants), Acta 1978. The legislation has the objective of affording protection to phytogenetic creations and allows the

19) Núcleo de Economia Industrial e da Tecnologia (2002) *Estudo de competitividade por cadeias integradas no Brasil: complexo da saúde*. Unicamp, Campinas.

'Cultivar Ownership Titles' issued to be managed by the Registro Nacional de la Propiedad de Cultivares (National Registry for Cultivars Property).

An interesting example of IP management occurs with the family business RELMÓ. The business strategy concentrates on the production and sale of plant seeds for mainstay crops: soybeans, wheat and maize. The growth of this company in recent years is essentially based on license agreements linked to IP. There have been important structural changes in the field of plant seeds. Companies operating vertically (ranging from the genetic enhancement of seeds to their subsequent sale) are declining while companies capable of coordinating licenses with other companies for new technologies and thus accelerating technological diffusion and increasing the flow of germplasms are arising. Within the scope of Argentine legislation on the matter of ownership of plant varieties, RELMÓ has been capable of negotiating licenses for their varieties and of marketing varieties created by third parties. Certain of these agreements stand out, such as with the Monsanto Corporation for the use of the Roundup Ready gene and the Bt (MON 810) gene. There are also agreements with other plant seed companies and public institutions. An example of this is an agreement for the genetic improvement of subtropical maize germplasm with the Argentine Instituto Nacional de Tecnologia Agropecuária (INTA; National Institute for Agriculture and Livestock Technology). This public–private partnership brings benefits to both institutions. The INTA provides the germplasms owned by them, installations and technical personnel, while RELMÓ pays the operational expenses. The hybrids obtained are marketed by RELMÓ, which maintains an agreement for exclusive exploitation with the option of granting licenses to third parties. RELMÓ pays the INTA royalties for the sale of the seeds.[20)]

In Chile, the main norms relating to IP are Law no. 19.996, of 2005, for industrial property and Law no. 19.342, of 1994, concerning the Rights of New Plant Variety Obtainers.

It should be stressed that Chile maintains free trade agreements with several countries, among which: Mexico (1997), Canada (1997), the European Union (2002), South Korea (2003), the European Free Trade Association (2003) and the United States (2003). Overall, these agreements contain specific provisions in the field of IP.

A scant 21 patents were granted Chilean inventors by the PTO in 2004. This was not very different from the previous years: 18 patents (2000), 20 patents (2001), 18 patents (2002) and 16 patents (2003). The numbers relating to PCT patent applications were (see Table 6.2): one application (2000), three applications (2001), six applications (2002), nine applications (2003) and six applications (2004).

According to Santibáñez[13)] (2003), Chile does produce knowledge and technology in the university–corporate sphere, but remains unable to promote the trans-

20) Domingo, O.A. (2004) IP management in the development of a medium-sized Argentinean seed company. *WIPO SMEs Newsletter*, **June**, http://www.wipo.int/sme/en/case_studies/relmo.htm.

fer of technology in a satisfactory manner. According to estimates for 2001 advanced by the author, the projects supported by the Fondo de Fomento al Desarrollo Científico y Tecnológico (Fund for Fostering Scientific and Technological Development) generated: 284 new products, 72 improved products, 100 new processes, 151 improved processes, 104 new services, 17 improved services and 139 other results. Of the total 868 results obtained, 17% were liable to be patented. However, this only gave rise to six patents and 12 patent applications.

A trait common to Latin-America is the poor institutional capacity for managing IP. Capacity building programs directed at IP for biotechnology (identification of patentable inventions, patent drafting, and negotiation and marketing of patents, trademarks and other intangible assets) should be considered as priorities by governments so as to promote correct appropriation of the results generated from local research and take advantage of partnership opportunities at local or international levels.

Low patenting activity together with high rates of patenting by non-residents in these Latin-American countries has caused an increasing dependency on foreign generated technologies.

6.5 Conclusions and Final Remarks

As can be seen, the innovation systems of Brazil, Argentina and Chile are dissimilar, but remain marked by poor articulation and by the frailty of certain cogs within the mechanism. These systems inherited the policies and practices inherent to development models that did not prove successful and that did not place value in technological innovation, or partnerships and appropriation strategies. The technological accumulation of these countries, much as that of other Latin-American countries, occurred despite little investments in leading edge technology and many projects that privileged adapting and improving the practices common to other countries.

Human resources were prepared and funds were invested in basic research, but with little effort to link the costs of research to corporate interests. There were also few efforts to increase competitiveness through international partnerships.

The 1990s were marked by freer trade and attention to quality and technological upgrades. The purchase of machinery and equipment seemed to fulfill the government innovation policies still strongly ingrained with a linear perception of the innovation model. Conceptions of the true complexity of this innovation model only began to be incorporated into certain innovation policies in Latin-America in the late 1990s.

21) Barbosa, A.L.F. (1999) *Sobre a Propriedade do Trabalho Intelectual: Uma Perspectiva Critica*, Editora UFRJ, Rio de Janeiro.

22) Buainain, A.C., Carvalho, S.M.P. and Carvalho, S.P. (2000) Propriedade intelectual em um mundo globalizado, in *O futuro da indústria: Cadeias Produtivas* (ed. J.M. Silveira), PUBLSHER?, Brasília.

After the implementation of the TRIPS Agreement, the greatest challenge today is how to accelerate the technological progress of these countries and how to promote partnerships, as much within the network as those for the transfer of technology involving both north–south transfers and those within the southern hemisphere. Certain aspects of the knowledge generation and exploitation processes must be considered: the processes are not linear; the search for new products, processes and purposes often fails. The involvement of Latin-American corporations with the R&D activities of universities and research institutes is another critical point, as well as a research schedule, which must be intended for technology of medium and high aggregate value, with emphasis on seeking solutions for local problems. The intake of researchers by the corporations is one of the challenges faced by innovation fostering policies.

Overall, the positive characteristics present throughout Latin-America are the excellent quality of the human resources and regulatory limits (IP and biosafety) fully in accord with the TRIPS Agreement, providing an environment of legal certainty to long term investments in biotechnology. Other factors that should be taken into account during negotiations with Latin-American countries is government purchasing power and local market size, with both factors being most interesting for the commercial exploitation of biotechnological creations. With regions notably rich in biodiversity, Latin-America garners favorable conditions for competitiveness in the field of biotechnology.

The perspectives with investments in biotechnology innovation is for the generation of wealth as well as other benefits, such as the development of specific therapies and medicines for the diseases that afflict the region, the development of agribusiness and the reduction of expenditure with machinery and equipment imports.

The dynamic for managing and generating biotechnology innovation is notably interdisciplinary, oriented by the efforts of doctors, biologists, agronomists, engineers, administrators, chemical engineers, pharmacologists and even consumers. It calls for constant articulation between the research bases and both the private and public industrial sectors as well as permanent investments in the postgraduate system. Building up capacity for the protection and commercial exploitation of IP is crucial. For Latin-American countries, the assimilation, coordination and regulation of so many different aspects is a complex challenge, but well within the realm of possibility.

Part II
Perspectives from Different Participants/Players

Technology Transfer in Biotechnology. A Global Perspective.
Edited by Prabuddha Ganguli, Rita Khanna, and Ben Prickril

ISBN: 978-3-527-31645-8

7
Technology Transfer in Agricultural Biotechnology: Impact of Applied Plant Sciences on Industry

Marc Cornelissen and Michiel M. van Lookeren Campagne

7.1
Introduction

The past three decades have shown considerable advances in plant science. This greatly enhanced our insights of metabolic pathways and plant development. In addition, it led to advances in enabling technologies. Together, these developments enormously increased the potential for the usage of plants as a sustainable production system.

Foreseeing the range of opportunities, the producers of agricultural chemicals already embraced plant science in an early stage of the developments. The introduction in crops of herbicide tolerance and insect control (the so-called input traits) was the first objective of the newborn agricultural biotechnology industry. At that time the opinion among industry leaders was that they could discover and develop new traits independently.

This opinion is now rapidly changing. The latest advances in plant science have opened the door for the application of crops as a renewable resource for a wide array of plant-based specialty products. Among these are bio-fuels, nutraceuticals, industrial raw materials and plant-made pharmaceuticals.

The main representatives of the agricultural biotechnology industry are already actively pursuing the new opportunities. The investments and timelines necessary for the development of these scientifically complex agricultural products has brought the insight that more intense and more directed interplay between the science community and the industry is essential for future innovations. The industry has great expectations of the intensification of this public–private cooperation as it holds the promise of sustainable solutions to many of the current worldwide unmet needs.

In this chapter the developments in agricultural biotechnology are first placed in an historical perspective. We then look at the developments in the market and the rising need for technology transfer through open innovation platforms. We also look at the opportunities this will offer for the various partners in the value chain. This is followed by a discussion of the specific characteristics of technology

Technology Transfer in Biotechnology. A Global Perspective.
Edited by Prabuddha Ganguli, Rita Khanna, and Ben Prickril

ISBN: 978-3-527-31645-8

transfer in agricultural biotechnology. Owing to their special role in stimulating technology transfer we then identify some challenges specific for start-up companies in this sector. Finally, we take a brief look at the future and the impact the transition to a knowledge based bio-economy can have on governmental budgets and our society.

7.2 Historical Perspective

7.2.1 First-Generation Crops

The first scientific breakthrough relevant for the industrial involvement in this field occurred around 1980. It was the possibility to bring foreign genes to expression in plants with the help of DNA transfer through the plant-infecting bacterium *Agrobacterium tumefaciens*. Using this system, the first traits to be introduced in crops were herbicide tolerance and insect control. That led to the chemical industry embracing the technology, and acquiring start-up companies founded by the pioneers of these techniques and seed companies to secure the vehicle to bring this technology to the market. Around 1990 the first generation of these innovated crops began to take shape in field tests.

The first modifications of crop plants with these genomics and recombinant DNA techniques in the 1980s can now safely be regarded as a breakthrough innovation. Since then a complete new industry evolved around these techniques: the agricultural biotechnology industry. In the process it reshaped the plant breeding process – an important economic activity that has a history of many centuries. As to be expected, many hurdles had therefore to be overcome before the first commercial agricultural biotechnology products were introduced in the field in 1996. Ever since then the market for these products has been growing at 10–15% per year (see Figure 7.1).

7.2.2 New Innovation Wave

At the end of 1980s, scientists discovered new ways of identifying traits at the DNA level, which opened the way for DNA marker-assisted breeding. This revolutionized the selection process as it enabled an efficient and cost-effective selection of desired and valuable characteristics.

At the beginning of the 1990s this marker technology was sufficiently developed to be adopted by industry. This development was further stimulated by scientific progresses in gene mapping in combination with the development of cheaper markers for the precise and rapid identification of specific genes in the plant genome.

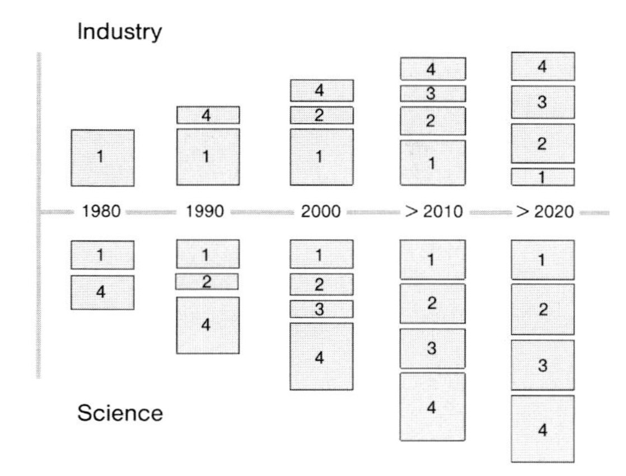

Figure 7.1 Relation between scientific breakthroughs and industrial product innovation. Ranking based on costs and impacts: (1) conventional breeding, (2) marker-assisted breeding, (3) breeding by design enabled by system functionality – mode of action knowledge and (4) predictive transgenesis – mode of action optimization.

At the same time it became clear that traits need to be embedded in an appropriate genetic environment to reach their full potential. In essence, the germplasm determines the performance of the trait. Thus, the combination between seed and biotechnology companies became essential. Especially outside Europe this notion spurred a further consolidation of the seed industry. Players able to integrate both these activities are now leaders in the sector. As soon as Europe embraces the innovative crops generated with these new technologies it may be expected that this consolidation will accelerate in this region as well.

The increases in understanding also led to new breeding methods. The implementation of these so-called advanced breeding techniques allowed non-genetically modified organism (GMO) solutions for high-impact differentiating traits. This allowed the industry to replace a subset of GMO solutions by non-GMO alternatives. All agricultural biotechnology industry now prefer non-GMO solutions as these are more cost-effective to develop by not requiring a heavy deregulation process. The ability of agricultural biotechnology to develop high value, differentiating GMO and non-GMO traits, triggered the industry to introduce the term 'high impact traits' (HITs).

7.2.3 Third Innovation Wave

Around the turn of the century, scientists published the sequence of the full Arabidopsis and rice genome. In combination with extensive analysis of mutants and the integration of knowledge of several plant science disciplines, this led to a greatly enhanced understanding of metabolic and developmental pathways. As a result it became possible to redesign metabolic pathways. This in turn opened the door to creating new functionalities. Almost overnight the innovative scope for the agricultural biotechnology industry broadened enormously. Apart from the optimization of existing ones and the redesigning of less-optimal solutions obtained by classical breeding strategies, this also enables the creation of crops for many new applications, like bio-fuels, nutraceuticals and plant-made pharmaceuticals. In the coming two to three decades countless new opportunities will follow and many of these will be based on new advanced plant science knowledge.

7.3
Market Development

7.3.1
Overwhelming Success

The introduction of the first commercial GMO crop in 1996 was the beginning of an overwhelming success. Also, thanks to the later introductions of HIT-carrying crops, HITs, the market has since grown with at least 10% per year. In 2006, the global area of these crops surmounted 100 million hectares planted by more than 10 million farmers. Cropnosis (www.cropnosis.com) estimated the global market value of GMO crops alone in 2006 at € 4.7 billion. This represents 16% of the € 29.6 billion global crop protection market and 21% of the approximately € 23 billion global commercial seed market in 2006 (see Figure 7.2).

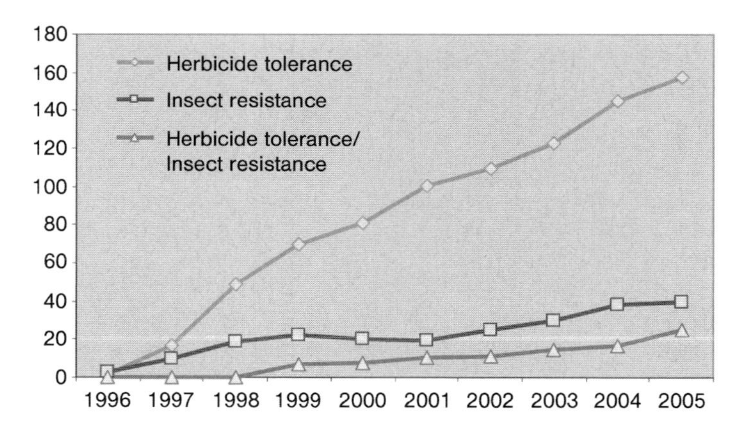

Figure 7.2 Global area (million acres) of biotechnology crops, 1996–2005, by trait. Source: Clive James (2005).

7.3.2
Full Potential Not Reached

In spite of this success, the full potential of the new traits and crops as forecasted by industry has not been reached. This is due to three factors:

1. Technological complexity. As is the case with the introduction of all new breakthrough technologies, the complexity of the development and introduction of new traits was at first underestimated. Soon it became apparent that gene sequencing and expression profiling were not the panacea for new products. The flood of gene leads created a bottleneck in lead verification. As in the pharmaceutical industry, the agricultural biotechnology industry soon started to realize that the conversion of leads into real products was much slower and more expensive than first estimated. This led to the genomics bust at the end of the 1990s.
2. Deregulation process. From the beginning onwards the development and market introduction of new GMO traits is surrounded by an intensifying deregulation process. As a result the estimated R&D costs involved in meeting all the regulatory requirements increased about 5-fold to about € 25 million in 2006. Most importantly timelines for product introduction were also increased by a minimum of 2 years, which negatively impacts the economic value. This now forms a large entry barrier in the agricultural biotechnology market. It also limits the range of traits that qualify for commercialization. Presently, only potential blockbusters developed by companies with a global presence now stand a chance of reaching the market.
3. Lack of embracement of GMO technology in Europe. The more or less general anti-GMO climate in Europe had many effects. First of all it restricted the potential planted area significantly. It also impeded the application of this technology to heavily traded crops as wheat and rice, as the lack of acceptance in Europe would cause unmanageable conflicts in the normal trade and processing channels. It also generated a climate in which politicians, downstream industry and society in general became highly skeptical about the added value of this type of innovation, and became reluctant to associate themselves with it.

7.3.3
Future Developments

The ability to create crops with new functionalities and their broad applications as sustainable production systems have increased the market potential for plant-based products enormously. The sector already expects to realize more than a 5-fold increase in the demand for plant-based specialty products in the next two decades. Underlying this increase is the prospect for plant products also finding applications outside their traditional value chains, tapping into an estimated end market of over € 0.5 trillion. These developments are popularly referred to as the transition to the knowledge-based bio-economy.

All stakeholders, ranging from academia to consumers, will benefit in a sustainable way from the development of these opportunities that in turn will contribute to regional welfare, job creation and consumer benefit. Already these prospects have led to significant increases in investments in R&D by the agricultural biotechnology industry and the public sector. In addition it is also opening up many new opportunities for start-up companies and funding opportunities in focused fields for academia.

7.4 Technology Transfer Through Open Innovation Platforms as a Key to Success

7.4.1 Introduction

A requirement for an innovation to reach the market is that it creates added value to the consumer. Plant-based innovations find their customers at increasingly downstream points in the value chain. As a consequence, a key success factor will be the transfer of technology as an added value across the value chain. This must go hand in hand with the development of new business models. Reciprocally, the unmet needs of the customer need to be transferred backwards through open innovation over several steps in the value chain to be translated into research goals for the plant biotechnologists. Such transparency and cooperation across the value chain are presently seen in several other industries. In many respects the pharmaceutical industry will provide a model for the future working principles across the agricultural biotechnology value chain.

7.4.2 Need for Innovative Raw Materials as a Driving Force

For the end users of plant-based products, differentiation through new products with new functionalities is essential. This directly affects their ability to at least maintain, but better still increase their competitive edge, and thus market share and profit margins. The present end users of plant-based products have been faced with a slowing down of breakthrough innovation through processing automation. This is directly related to the maturity of the technology. Consequently the focus is now shifting to innovations in raw materials and plant based specialty products (see Figure 7.3).

7.4.3 Intensification of Coordination and Communication

The development of sustainably produced novel, HIT-derived raw materials and plant-based (specialty) products requires an intensification of the communication and coordination in the value chain. This is essential to the creation of the neces-

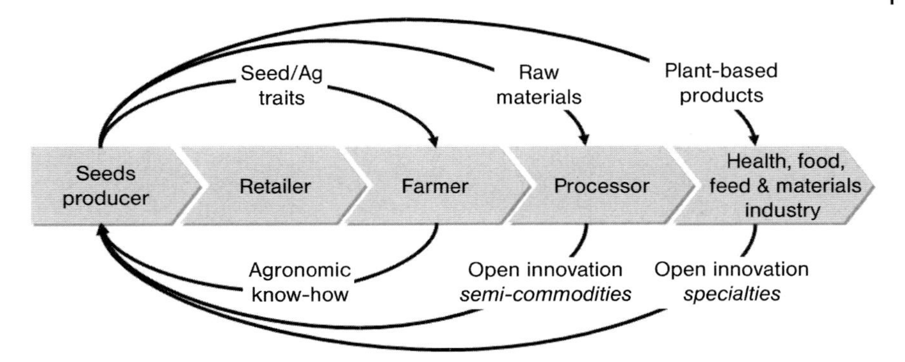

Figure 7.3 Flows of communication necessary to obtain the necessary transparency and coordination in the value chain for a win/win situation to occur during the creation of new HITs.

sary win/win situation that provides all parties involved with a balanced share of risks and revenues.

In practice, the end users and processors must therefore share their unmet needs backwards into the value chain. There they have to be translated by the agricultural biotechnology industry into new HITs. The specifications of these need to be communicated forward into the value chain in reiterative cycles of testing and re-evaluation. Specific for agricultural biotechnology is that this whole process is a high-risk undertaking as the developmental timelines are long (about 13 years) – a characteristic the consumer-oriented industry is not accustomed to. This places high demands on the quality of the communication and coordination efforts inside the value chain.

7.4.4
Involvement of Society

The success of the resulting innovations will depend on a fluid passage through the value chain. In many cases this also requires consumer acceptance in the end market before the introduction of the novel products. This requires the involvement of representatives of the society such as consumer organizations, politicians and regulators in an early stage of the development of a new HIT.

In this pre-market preparation politicians are of special interest. They have the power to create and define the borders of the markets for new HITs, as they have done in the past. A relevant example of this is the recent goal set by the European Union for the replacement of fossil fuels by bio fuels. Since the yields of present bio fuel production systems are in general too low for a sustainable commercial development, it opens a market for HITs with the improved yields of bio-fuels required to reach these targets. It also shows the importance of the dialogue between politicians and the industry as reaching these sort of targets set out in the political process depends on the ability of the industry to develop suitable HITs.

7.4.5
Responsible Industry

The past decades have already seen an increasingly close cooperation between the industry and regulators resulting in a growing atmosphere of trust. As the complexity of the new HITs increase this cooperation will further develop. Central to this is an agricultural biotechnology industry as a responsible initiator and developer of novel plant-based products and their markets. The industry is fully aware of the potential for adverse effects that new HITs may pose to the environment and the society in general. Also, the industry realizes that a general atmosphere of trust surrounding these innovative products is essential for successful market introductions of new innovations.

From the beginning, the industry therefore developed a high level of self-control and self-regulation. As a result, in many countries the trust of consumers and their representatives in the industry is also growing. This is paving the way for a new win/win situation. It can occur when the cooperation between industry, public and regulators reaches a level of understanding at which the industry is allowed to contribute to a unification of deregulation processes across the various regions.

Industry as developers of new HITs and their applications have the best insights in possible adverse effects. Their contributions will therefore increase the quality of the deregulation process. At the same time a leveling of the regional differences in the deregulation process will decrease its complexity. This greatly helps to limit the costs and the timelines of the R&D process of new HITs. The possibilities for the development of new HITs will then increase, which will greatly stimulate the transition to the bio-economy. Of course contributions of the industry or perhaps even an industry-driven deregulation process can only be realistically accomplished when it is safeguarded by a structured input of the various governmental and political bodies. It also requires a careful control and sanctioning process by the various governments, as well as a carefully designed independent inspection process.

7.4.6
Common Interests

As discussed the society and the agricultural biotechnology industry share many interests, which perhaps presently are not fully recognized by all. When this realization gains a better foothold, also more far-reaching forms of public–private cooperation become possible which offer various new win/win opportunities. One of them is, for instance, the advantages offered by more direction and cooperation in the private and public research efforts. At this moment this direction is rather limited. This leads to various sorts of inefficiencies in the research efforts. On the side of the academia this of course translates to a less-efficient application of public money.

A discussion between academia and the industry aimed at common goal setting for their plant-based research can enhance the efficiency and increase the pace of advances in plant science. Academia will profit from it as it will enhance their status as centers of excellence for applicable research. There are many examples in other sectors in which this has led to a greater access to students and funding.

More direction in the research efforts will also create more chances for start-up companies. Logically, the fundamental findings around which they are erected will appeal more to the industry as they are the result of a coordinated effort. More start-up companies will further enhance the funding of academia via participation and licensing fees. That will also contribute to the status of the institutions, so that a self-sustaining process is started.

Apart from more rapid and more directed advances in plant science this win/win situation also offers other advantages to the industry. For instance, it offers the prospect of sharing risks and the access to more and more readily applicable innovations generated by the start-up companies. It also holds the promise of future employees with educational profiles better suited to the needs and practices of the industry.

7.4.7
Open Innovations Platforms

These developments will increase the rate and importance of transfer of technology in agricultural biotechnology enormously. A special challenge is to develop a system that applies the necessary direction to the innovation process required for the increasingly complex HITs. It will require a high level of structuring as any partners and alliances will soon be involved. In other sectors this has led to the development of open innovation platforms. A great advantage of an open innovation platform is that every participant can concentrate on their strong points. This sharing of forces is a very profitable method of cooperation. It rewards every participant much more than they could ever hope to achieve on their own. For instance, it creates the possibility to benefit from many applications of a finding or developed technology in different markets, rather than to be limited to the income generated by the ones a given company can successfully develop independently. The structured cooperation in an open innovation platform therefore lowers the risks for all participants while simultaneously optimizing revenues and profits.

7.4.8
Coordinator Role in Industry

Central to an open innovation platform is a coordinator or system integrator. They must have the expertise and means to investigate the unmet needs of processors and end users, and be able to translate these in criteria for new traits and functionalities. They also must be able to bring together, and integrate all necessary innovations produced by the various partners into a successful an new trait and market the resulting HIT crops worldwide. In addition, they of course must also be able to maintain the dialogue with all organizations, regulators and politicians in many countries simultaneously, as well as be able to professionally manage all other relations, alliances and license contracts involved.

This role fits best with the capacities and abilities of the present agricultural biotechnology leaders. The transition to this new role will nevertheless mean a great change for them as the agricultural biotechnology companies are presently mainly focused on the role of independent developer and breeder of crops. Held against the present practices within the agricultural value chain and the agricultural biotechnology industry, this new way of organizing the development of innovative HIT crops may also imply that the roles of other partners in the value chain slightly change.

7.4.9 Opportunities for Academia and Start-Ups

An open innovation platform also tends to make it easier for academia and start-up companies to generate income from their expertise and findings, and thus they increase their opportunities. Partly this is because it provides them with easy access to licensees. The structuring of the innovation process also increases the insights in which findings and new developed technologies may be of value and which not. This provides both academia and start-ups with a well-documented opportunity to consider a specialization on certain research fields. Presently, logical fields of interest are, for instance, the various aspects of metabolic pathways and their regulation.

7.4.10 Key Factor: Technology Transfer

All these developments and opportunities will only materialize if a good reward system is developed for the transfer of agricultural biotechnology technology. Only then will the development of HIT crops remain an economically viable activity and will it appeal to investors. Systems for the transfer of technology developed in other sectors like the pharmaceutical sector can serve as a model, but cannot be copied directly. Some adaptations are necessary to deal with the specific characteristics of the R&D process and the development of HIT crops.

7.5 Specifics of Technology Transfer in Agricultural Biotechnology and Their Consequences

7.5.1 Introduction

There are several specific characteristics of the developmental process and HITs themselves that make the innovation process in agricultural biotechnology unique. At present, these features put a brake on technology transfer and the development of new HITs. They need to be addressed in the near future to establish the technology transfer necessary for successful open innovation platforms and thus for the transition to the bio-economy. This poses serious challenges.

7.5.2 Transition Points

The transfer of technology in the developmental process and market introduction of new HITs involves a series of value transition points. Some of these share a feature that needs to be addressed to stimulate technology transfer and these are discussed in the following subsections.

7.5.2.1 Transfer to Agricultural Biotechnology

This concerns the technology transfer from academia to start-up company, from academia to agricultural biotechnology leaders and from start-up company to agricultural biotechnology leaders.

Representative of the agricultural biotechnology leaders in past negotiations experienced that academia and start-up companies often have exaggerated expectations of the value of their findings. This originates in an often limited knowledge of the industrial R&D process, associated drop-out rates and costs of HITs. This is not very surprising as the industry for many years hesitated to provide insight in the cost and revenue structure of their R&D process, as this affected their position in the negotiations. Unfortunately, negotiations frequently failed due to unbridgeable gaps in discussions on milestones and royalties. Clearly bridging these gaps calls for more trust and more open communication between the industry and these parties. Once the industry realized this they started to provide verifiable information about the cost and revenue structure of their R&D processes. The increased transparency has started to clear up the atmosphere, but further efforts in this field are certainly needed. By nature the academia and beginning start-up companies lack commercial experience. At the same time, this is badly needed to fully comprehend the processes in the agricultural biotechnology value chain, and the implications of a commercial undertaking that has a developmental timeline of 13 years, a success rate of less than 10% and an investment profile in which about 50% of costs have to be made in the early phases of development (see Figure 7.4).

7.5.2.2 Downstream from Agricultural Biotechnology

This concerns the subsequent technology transfer from agricultural biotechnology industry downstream in the value chain to retailers, from them to farmers and then onwards to the processors.

An important consideration in this technology transfer is the switching costs. These are the specific costs involved in the required adaptations in machinery, techniques and processes in the downstream partners that can be directly related to changes in characteristics of the innovative crop in comparison to the former

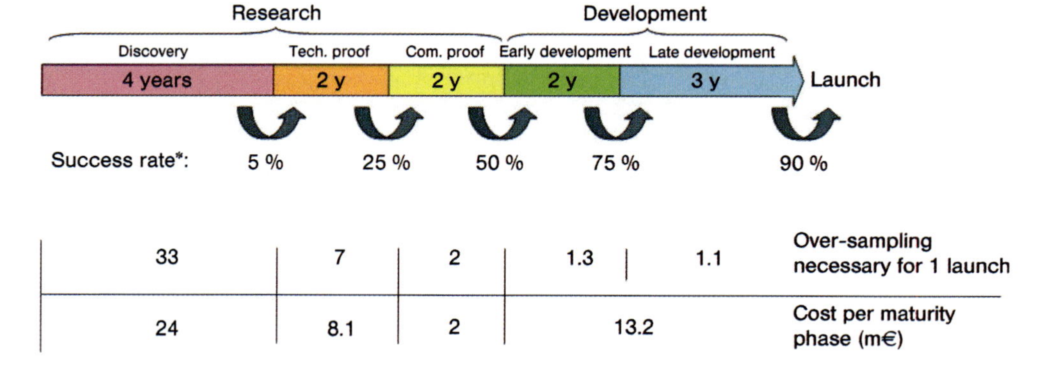

Figure 7.4 Timeline, costs and necessary number of crop ideas required in the various developmental stages needed for the successful introduction of one new top crop.

ones. The experience with the first GMO crops and HITs showed that every new trait will result in smaller or bigger changes in characteristics. These range from the need for new farm equipment due to different plant architecture to the need for identity preservation of the improved products. Downstream switching costs are therefore unavoidable and of course have to be covered by the added value of the new HIT.

Determining the switching costs increases the R&D costs, moreover so as it frequently requires an intensive testing and evaluation process with the downstream partners. This contributes to a situation in which only HITs with very high potential can meet the determined added-value criteria, thereby significantly limiting the opportunities for HITs. To the frustration of academia it also influences the value the industry puts on new traits that bring high switching costs with them. The switching costs therefore also put a brake on technology transfer. The sector itself can do little to remove these limitations. This therefore provides a challenge for others like regulators and politicians.

7.5.2.3 To End-Consumers

As in all other chains, the last step in the agricultural biotechnology value chain, from processor to end-consumers, determines the success of a new HIT. On the one hand, the long developmental timelines in agricultural biotechnology provide ample time for pre-market preparation. On the other hand, this also makes it a high risk process as the acceptance of many HIT-based/containing products by end-consumers relies to a large part on emotions. Incidents, sudden and unpredictable by definition, can therefore cause major shifts in the acceptance of a HIT-based product. More than in any other sector, the success of technology transfer in agricultural biotechnology, as well as the success of new HITs, is therefore directly tied to leaders in society embracing and supporting the technology and the products. It is a challenge for the whole sector to gain this support.

7.5.3
Expression of Traits

Another important specific characteristic in agricultural biotechnology is that the technology itself is not the product. Every new trait is bred into a broad array of different varieties. It must come to expression and lead to similar results in different genetic backgrounds. The practical applicability of the trait clearly depends on this and has therefore to be recognized in the reward structure. It also implies that a trait to be licensed needs to be introgressed and tested in proprietary germplasms of the licensee at an early stage. Unfortunately, this is a multi-year process. This is further complicated by the need to prevent the further lengthening of the already long developmental timelines. Therefore, new traits have to be incorporated and tested in many different germplasms simultaneously. This is of course more costly and involves more risks than testing one after each other. The industry therefore tends to strive for risk and cost sharing during this test phase. Together, these two features form an important entry barrier for trait licensing as logically companies normally only want to license technology with proof-of-concept in the commercial crop of interest.

7.6
Technology Transfer and Start-Up Companies in Agricultural Biotechnology

7.6.1
Introduction

The dedication of their leaders and the flexibility with which they can operate make start-up companies especially suited for the generation of innovations. In many sectors the majority of innovations and subsequent transfer of technology can be related to these young and dynamic companies. However, in the 1990s the agricultural biotechnology leaders showed little compassion for start-up companies: they could do without them. As described above, in recent years this viewpoint has changed completely. Nowadays, the agricultural biotechnology industry regards start-up companies as essential for the further development of innovative HITs. In principle there are now already ample opportunities for start-up companies. The specifics of agricultural biotechnology make the conditions for start-up companies less favorable. Some of these issues need to be addressed to create a stimulatory environment necessary for the emergence of significant numbers of new start-up companies. This requires a combined effort (see Figure 7.5).

7.6.2
Less-Favorable Prospects

Despite the great array of product and specialization opportunities, the prospects for start-up companies in agricultural biotechnology are still not very favorable.

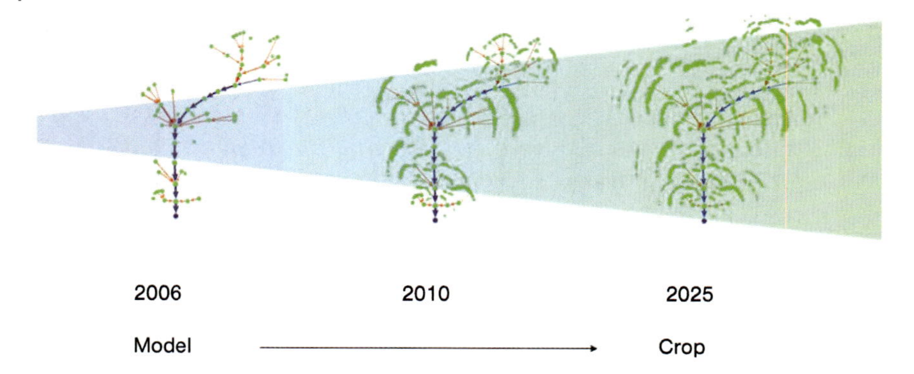

Figure 7.5 Increases in knowledge and complexity need to be captured during the development of a HIT from the first laboratory model to the manifestation of the crop in the field.

This is due to the characteristics of HITs as well as the agricultural biotechnology sector in general:

1. Poor translation of laboratory ideas. As described, the transition of laboratory ideas to commercial traits is poor. This resulted in a licensing process that involves long and costly pre-testing of the expression of traits. Also, it made the industry leaders strive for risk sharing. This significantly limits the chances for the normally cash-craving start-up companies.
2. Long developmental timelines. The main income of start-up companies in agricultural biotechnology will depend on royalties realized after market introduction of the HIT in which their findings are incorporated. The developmental timelines are about 13 years on average. A start-up company will therefore have to bridge an unusually long period on limited financial possibilities or find other sources of income.
3. Limited number of clients. The chances of start-up companies are further limited by the reality that there are only a handful of companies to sell products to. This implies a large dependency of the start-up on their clients. In all sectors this is a very fragile situation with a high chance of failure.
4. Lack of benchmarks. Since it is also a completely new technology field, it is also very difficult to put a price tag on the transfer of findings and new developed traits. There simply are no benchmarks that can be used in the calculation of values, royalties and so on.
5. Stacking of traits. As the number of stacked genes in a HIT increases, the income to be generated by an individual trait decreases. This poses a threat to the future revenues of start-up companies.
6. Absence of entrepreneurs. In many western countries there is a increasing shortage of labor. In particular, the rare, highly educated individuals with entrepreneurial skills nowadays have many other options. As long as the prospects in agricultural biotechnology remain low, they will find their future elsewhere.

7. Willingness of investors. As already stated, new start-ups will only take flight if investors are willing to fund the developments. This is equally true for the R&D within the existing multinationals and other companies. Although the transition to the bio-economy provides interesting prospects, the risks involved, earn-back times and return on investments in agricultural biotechnology dampen the willingness of the majority of independent investors to provide funds.

7.6.3 Stimulatory Measures

Presently only a few companies exist that can be recognized as an independent agricultural biotechnology start-up and most of these have met with limited success. At the same time, start-up companies have a crucial role to play in the development and transfer of technology necessary for the transition to the bio-economy. It is in the interest of the sector as a whole as well as everybody else embracing these developments to stimulate the founding of new start-ups. A stimulatory environment for start-up companies can only be created by addressing the specific factors that currently limit their prospects. Clearly, this should be the result of a combined effort of the industry, academia, governmental bodies and politicians.

7.7 Future

7.7.1 Introduction

In the coming two decades the ongoing developments in plant science and the introductions of new HIT crops will cause an evolutionary change in the agricultural biotechnology sector. At the same time the transition to the bio-economy will boost the agricultural sector. Since this transition, for instance, promises to decrease our dependency on fossil energy, also the environment will benefit greatly. The economies of the countries that embrace these developments as well as their populations will also benefit in other ways from this evolution.

7.7.2 Diversification and Specialization

The trend towards open innovation platforms is already driving an evolutionary process in the agricultural biotechnology sector towards specialization and diversification. As described above, the future of agricultural biotechnology is in the plant-based production of specialty products. The field of possibilities for these traits and functionalities is, however, so broad that it cannot be covered by an individual company. Logically, the present industry leaders will therefore specialize in specific areas. This specialization will be accompanied by a diversification.

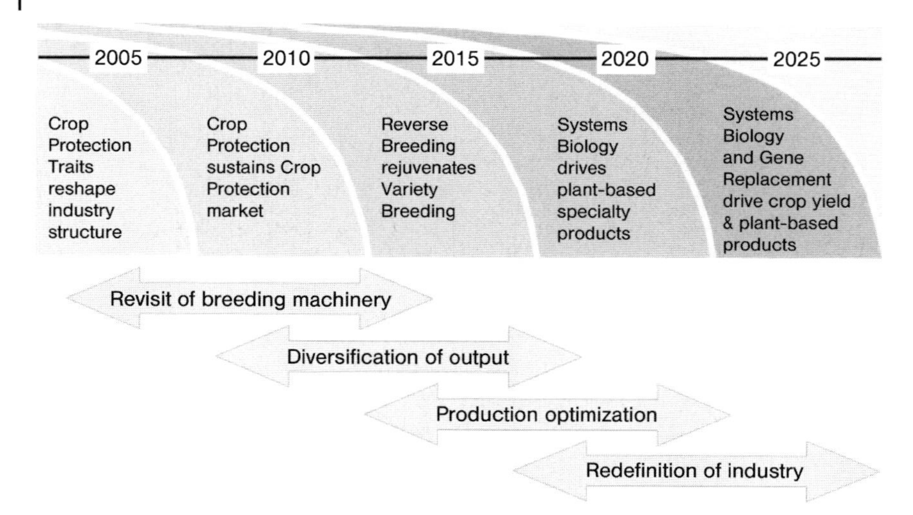

Figure 7.6 Survey of the evolutionary steps that will take place in agricultural biotechnology during the transition to the bio-economy in relation to the changes in the orientation of the R&D involved in the realization of new top crops.

7.7.3
Evolution in R&D

In the years to come, the industry will first use the new insights in plant science and the newly developed techniques to revisit the past products of classical breeding, as well as the first attempts to develop GMO crops. Once this supply of acknowledged ideas is finished, the industry will start to endeavor new HIT crops in full. Logically, this will start with relatively simple ideas, and then evolve to more ingenious and more complex solutions. In particular, the development of the more complex HIT crops requires plant system knowledge comprising the interaction of several hundreds of genes, proteins and various metabolic pathways. Since the understanding of the workings of one gene was estimated to require about 25 man-years, the investments involved in the successful development of these new HIT crops are too high to be carried by a single company. In this stage the open innovation platforms are therefore badly needed (see Figure 7.6).

7.7.4
Shifts in National Budgets

The first to benefit from the ongoing introduction of HIT crops and the resulting transition to the bio-economy is the agricultural sector. It is foreseeable that the prosperity in this sector will take such forms that the present level of agricultural subsidies can at some stage be lowered. That in turn creates the possibility for major shifts in the spending of national budgets. For example, about 43% of the

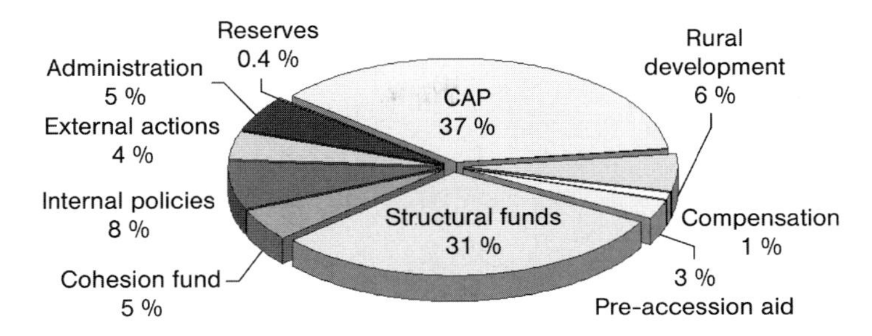

Figure 7.7 European Union expenditure 2004. CAP = common

budget of the European Union is presently reserved for common agricultural policies. On a yearly basis it spends about € 43 billion on agricultural subsidies and related issues. If Europe embraces the innovative traits and resulting HIT crops it is foreseeable that at least a part of this budget can at some future point be dedicated to other causes. It can, for instance, be applied to further spur innovations in agricultural biotechnology and other fields or it can be invested in education, health and infrastructure (see Figure 7.7).

7.7.5
The 'Stimulating Government'

It therefore seems that governments and societies have much to gain from the transition to the bio-economy. On a regional level the regulators and politicians can contribute to the acceleration of the transition process in various ways. This can, for instance, be achieved through investments in the education systems, R&D-oriented subsidies and the removal of unnecessary complexities in legislation and safety guidelines. The creation of new markets by enabling legislation can of course be an important stimulant of the transition. In this line of thought, for instance, fits legislation that promotes green crops that contribute to the national health.

One example is the legislation introduced in New York in the fall of 2006. This prescribes the use of healthy, unsaturated fats in the preparation of food by all restaurants and canteens. The introduction of it shows a close relation with the development of new traits and functionalities. Legal enforcement of these measures is only possible when there is an economic viable source of the required product. In this case it is the canola oil harvested from a HIT crop recently introduced in Canada that has high oleic/low linoleic oil profiles to decrease trans-fat formation during frying in cooking oil.

7.7.6
Self-Sustaining Process

The introduction of these renewable, health-promoting products and support by legislation will eventually contribute to a better environment and a healthier population in a given country. As a result, the costs of healthcare may be partially mitigated. If this becomes a reality, another shift in the budgets can be made. The transition to the bio-economy therefore holds many promises for industry, governments and society alike. The challenge is mainly to accelerate the transition to the bio-economy as much as possible.

8
Technology Transfer Issues in Biotechnology: The Industry Point of View

Florent Gros

8.1
Introduction

The need for novel medicines will rise as the population ages. The United Nations Population Division reports that in North America, 16% of the population was over 60 years of age in 2000. By the year 2025, it projects this percentage will rise to 25%; in Europe, the corresponding figure will be even higher. The increased life expectancy is primarily a consequence of the tremendous advances we have made in medical knowledge. For example, over the last 40 years mortality as a consequence of hypertensive heart disease dropped by 67%, the death rate for gastrointestinal ulcers declined by 61%, and for emphysema by 31%, and infant mortality dropped by a staggering 80%.[1]

Novel medicines are increasingly in-licensed by pharmaceutical companies. The R&D investments for each large pharmaceutical company already exceeds US$ 2 billion each year. The licensing competition is intensively focused to capture innovation from universities or smaller pharmaceutical companies. For instance, about a third of the new molecules in clinical phase III are in-licensed and more than half the products with 2004 sales of US$ 501 million to US$ 1 billion were in-licensed.[2]

The therapeutic biotechnology products, also called 'large molecules' or 'biologics', are believed to take a significant part of the future products portfolio. The earlier biologics developed in the late 1980s, such as human growth hormone, are now out-spaced by the prospect of many new therapeutically effective biologics. For instance, antibodies have the formidable potential to selectively target diseases and their clinical development proved to be fast.[3] Another stimulating factor is that generics competition for biologics is not as acute as for small molecules. Not surprisingly then, the United States is the most active in R&D

1) 2004 letter of D. Vasella to Novartis shareholders.
2) Novartis benchmark data.
3) Baker, M. (2005) Upping the ante on antibodies. *Nature Biotechnology*, **23**, 1065–72.

Technology Transfer in Biotechnology. A Global Perspective.
Edited by Prabuddha Ganguli, Rita Khanna, and Ben Prickril

ISBN: 978-3-527-31645-8

for biologics and the proportion of in-licensed large molecule from large pharmaceutical corporations declines from the United States (around 30%) to Western Europe (around 20%) to Japan (below 10%). Also, over 80% of technology alliances are within the United States and European countries.[2)]

The R&D investments lead to the creation of intellectual property (IP), amongst which the patents are the major vector to pay for the innovation expenses, but also to secure freedom-to-operate and create leverage in future product and technology trades. The licensing of the necessary tools for discovering and developing novel therapies starts at the research level up to manufacturing. A vast number of technology transfers happen through R&D collaborations, patent licensing, drug in-licensing and even material transfer agreements (MTAs) (e.g. for biological targets, assays, cell lines and biomarkers).

The objective of this paper is not to discuss patent laws, but rather to provide the industry's perspective about how patents and other IP are traded for the advancement of novel therapies and therefore better healthcare services. To understand the practices and issues related in particular to the field of biotechnology, one has to understand the:

- Diversity of the legal frameworks.
- IP dangers.
- Industry out-licensing prospects.
- Industry in-licensing practices.
- Challenges related to biotechnology.
- Industry motivation in driving R&D investments.

8.2 Diversity of Legal Frameworks

The legal framework for technology transfers is highly complex, with different local and international laws. Large pharmaceutical companies, with multiple research centers in many different countries, have naturally developed global licensing practices aimed at best utilizing these frameworks. The local framework includes competition laws, licensing practices of universities and other institutions, data exclusivity, and patent-related regulations. The global framework includes the Trade-Related Aspects of Intellectual Property (TRIPS) Agreement, the Biodiversity Convention and the Organization for Economic Cooperation and Development (OECD) Guidelines for the Licensing of Genetic Inventions.

Competition laws (also called anti-trust laws) play a significant role in transactions, such as for patent licenses, R&D collaborations and product in-licensing. The rules are complex and vary from country to country, but the objective is simple – to regulate the market powers. Therefore, certain restrictions used to (overly) dominate pricing, distribution, manufacturing, and even the use of improvements can be sanctioned.

R&D collaborations aiming at developing a product that will create a new demand – therefore establishing a new market – are, however, unlikely to immediately raise competition concerns. The European Block Exemption for Technology Licenses[4] is a good example in which certain hardcore (unacceptable) restrictions are perfectly acceptable until partners are viewed as competitors, which may take years after the first commercialization of the traded product or technology. Therefore, it is not unusual to see in many R&D collaborations a grant-back clause to the licensor on the improvements made by the licensee because the partners are unlikely to rapidly become competitors. Such provision may, however, raise serious competition concerns between established competitors. The new challenge for the industry in Europe is to continuously monitor their market shares and the need to ensure that their agreements remain defendable at any time, if there is a serious risk that the market share thresholds will be exceeded.

The competition laws also play a critical role in merging and acquisitions. In May 2005, the Commission cleared the proposed acquisition of Hexal, a German producer of generic medicines, by the Swiss company Novartis. This resulted in the creation of the largest European producer of generic drugs. Clearance was given, subject to certain conditions. Serious competition concerns arose, however, in three markets: prescription calcitonins used for treating osteoporosis in Poland, over-the-counter topical anti-rheumatics in Germany and prescription anti-gout preparations in Denmark. In Poland, Novartis and Hexal had substantial market shares with respect to calcitonins. In Germany, the proposed merger would combine the leading branded product, Voltaren (Novartis), with the leading generic, Dislac (Hexal). In Denmark, the proposed merger would have created a strong market position for anti-gout preparations. Clearance for the acquisition was granted on the basis that the following commitments would be made:

- In Poland, Hexal's products in the calcitonins area had to be sold.
- In Germany, Hexal's product Dislac had to be sold.
- In Denmark, Hexal's anti-gout preparations business had to be sold.

Licensing practices of universities also significantly influence the technology transfers to industry. The 1980 Bayh–Dole Law in the United States mandates university and government researchers share in any licensing payments received by their institutions, typically about one-third. Despite the successes of the licensing offices of US institutions, it is not unusual to be confronted with inventors or licensing experts that have an inflated idea of the value of their inventions. Also, many US universities – quite often the most prestigious – have been inclined to adopt a 'take it or leave it' stance. Hence, they sometimes prevent consummation of a reasonable deal by their ambitions. For instance, the two leading US universities in technology out-licensing – Stanford and Massachusetts Institute of Technology – are known to make it difficult to start companies based on technology deriving from their laboratories. Harvard is thought by many venture capital firms to be even harder. The challenge for the universities is that they want to

4) EC Regulation 240/96.

maximize the revenues for their institution from any basic research later commercialized, but if they are too onerous then venture, biotechnology and pharmaceutical investors will not invest, and there will be no revenues derived, or less than would otherwise be the case.

By contrast, in much of Europe and Asia, the majority of universities are state institutions and have a sound argument against their administration for charging a fee to access their IP. The fact that some European countries allow government researchers to own the IP resulting from their work, e.g. Finland (unlike Germany and France), or appealing inventor remunerations for government researchers (e.g. Sweden, Germany and France), does not yet suffice to reach the level of US licensing activities.

Government funding on technology transfers from universities to the industry may also be significant. For instance, if the university research is funded by the US Government, the industry beneficiary has then to substantially manufacture in the United States any commercial product resulting from the funded research. It is obvious that this obligation may conflict with global business strategies of large pharmaceutical companies and consequently they tend to walk away from funds from the US Government.

The Framework Program for Research and Technological Development is the main instrument for funding research in Europe and offers major funding opportunities for the biomedical research community. The Sixth Framework Program (FP6) was launched in 2002, for a total budget of € 17.5 billion allocated over the 4-year period 2002 to 2006, representing 3.4% of the European Commission's total budget in 2002. In 2005, the Commission called to award € 533 million in health research in particular to exploit the potential of genome information.[5] A FP6 funding typically gathers several universities and one or more industry players. The participants will own the IP of the results and can define among themselves the commercial arrangement that suits best.

Patent laws are of paramount importance. The financial valuation of a technology takes into account many variables such as, for example, the role of proprietary know-how, data exclusivity and patents. However, patents have a major influence.

When the value of patent rights appears uncertain (e.g. regarding infringement and/or validity) many US universities and start-ups have been inclined to adopt a pricing stance based simply on the mere potentials of patent litigation costs. For industry, the equation is a difficult one, being confronted with US litigation costs that will probably exceed US$ 1 million and with the uncertainty of proving the patents being invalid or non-infringed.

The *Merck v. Integra* decision on Safe Harbor[6] (relating to US experimental-use exemption) has a major influence on the licensing of biopharmaceutical research tools. It is apparent from the statutory text that the exemption from infringement in the United States extends to all uses of patented inventions that are reasonably

5) http://europa.eu/scadplus/leg/en/lvb/i23012.htm.

6) Legislation USC §271(e)1.

related to the development and submission of any information under the US Food and Drug Administration (FDA). This necessarily includes preclinical studies of patented compounds that are appropriate for submission to the FDA in the regulatory process. However, this Supreme Court decision has also created questions whether patent exemptions may, or may not, extend to very-early-stage drug discovery (i.e. to early drug screening). By contrast, the experimental-use exemption regulations in European countries and Japan are quite diverse, with less current controversy than in the United States (i.e. whether early drug screening and use of research tools would be exempted or not from patent infringement). In general, such research exemption is unlikely in ex-United States countries.

Data exclusivity is clearly separate from patent protection and can play a significant role in drug in-licensing where patent protection is not available or not sufficient to recoup the investment costs. It refers to regulations whereby, for a fixed period of time, drug regulatory authorities will not use the drug registration files of a pharmaceutical company to register a therapeutically equivalent generic version of the drug. The rationale is that where substantial cost has been incurred by the researcher, it would be unjust to deprive them of the legitimate and reasonable profits by allowing other persons to adopt its unfair commercial use. For instance, the information provided to the regulatory authority during the examination is the result of experiments and clinical trials spanning many years and costing significant amounts of money, up to US$ 800 million.[7] This practice has been endorsed in Article 39.3 of the TRIPS Agreement and implemented in a number of countries.

The Hatch–Waxman Act in the United States provides a 5-year data exclusivity for new molecular entities (NMEs), i.e. a period of exclusivity such that once a NME is approved, a generic version cannot be approved for 5 years. This Act also calls for a 3-year data exclusivity period for supplements requiring clinical trials. Other countries have similar durations (e.g. 6 years in China, 5 years in Australia and 5 years in New Zealand).

The new EU pharmaceutical legislation[8] has created a harmonized 8-year data exclusivity provision with an additional 2-year market exclusivity provision. This effective 10-year market exclusivity can be extended by an additional one year maximum if, during the first 8 years of those 10 years, the marketing authorization holder obtains an authorization for one or more new therapeutic indications which, during the scientific evaluation prior to their authorization, are held to bring a significant clinical benefit in comparison with existing therapies. This so-called 8 + 2 + 1 formula applies to new chemical entities.

International treaties also significantly influence technology transfer practices. The most important one comes from the General Agreement on Tariffs and Trade and the subsequent patent reforms made after the TRIPS Agreement. As a result, the global strengthening of patent protection still appears today as a major

7) Bandow, D. (2003) Demonizing drugmakers. The political assault on the pharmaceutical industry. *Policy Analysis*, **475** (May 8), 3–5.

8) Directive 2004/27/EC.

growth opportunities for all countries and for all kind of industries, whether pharmaceutical or generic companies.[9] On the other hand, the TRIPS Agreement also contains provisions for compulsory licenses (e.g. in the event of public health crisis), which shifts the balance of rights and obligations away form the patent holder.

The TRIPS conference in Doha (2001) was a precursor to many recent compulsory licensing debates. The Doha declaration indeed called for the creation of procedures for developing countries to implement compulsory licensing only when the need for drugs rises to a state of 'emergency'. The procedures set forth by the World Trade Organization specifically articulate mechanisms for industrialized countries to manufacture and export patented drugs pursuant to a compulsory license whereby the developing country does not have the manufacturing capacity to meet their demands internally, subject to safeguards, such as limited amounts of production so as to meet the demand, restrictions of parallel import and adequate compensations. More recently, on 24 June 2005, the Brazilian government enacted a legislation that would override Abbott Laboratories' patent on Kaletra, an anti-retroviral drug used to treat HIV and AIDS, and would enable generic production and sale of the drug nearly half the price that Abbott currently charges. In some instances, the debate for access to patented medicines is also acute in industrialized countries. For example, during the anthrax scare of 2001, the demand for Bayer's Cipro sharply increased and Bayer donated supplies of Cipro in Canada in response to compulsory licensing public threat for generic production.

The Convention on Biological Diversity (or 'Biodiversity') which came into force on 29 December 1993, is applicable in a large number of nations. The objectives of this convention are the conservation of biological diversity, the sustainable use of its components, and the fair and equitable sharing of benefits arising out of the exploitation of the genetic resources in the world. This Biodiversity convention was made primarily to avoid bio-piracy that could happen during bio-prospecting, bio-screening or transfer of genetic materials and does not call into question the patent system.

Of significance, pharmaceutical companies should obtain prior informed consent in a pro-active manner before gaining access to a country's biodiversity (e.g. either through local operators doing the actual screening or directly in negotiation with the local government). This a 'best practice' that the Swiss pharmaceutical companies consider to be of utmost importance. However, knowing that a governmental agency is competent to give prior informed consent and the extent to which organizations of indigenous peoples should be involved is extremely difficult.

Also, in any contractual arrangement there must be a benefit-sharing proportionality between those that take the risk of developing and financing new products and those delivering genetic materials. Although the R&D-based pharmaceu-

9) Nair, M.D. (2002) An industry in transition: the Indian pharmaceutical industry. *Journal of Intellectual Property Rights*, 7, 405–15.

tical sector does not heavily rely upon biodiversity resources from developing countries, the Biodiversity convention remains an important framework for all technology transfers concerning country-specific genetic resources.

The OECD Guidelines for the Licensing of Genetic Inventions,[10] adopted in 2006, constitutes a new framework that may further influence licensing practices. These guidelines are not intended to cover exhaustively all aspects of licensing practices and are not directly binding for the industry. However, they set out important principles which the aim to avoid licensors overly restricting access and use of genetic information or materials for those in the business of human healthcare.

In the venture of pharmaceutical companies to obtain the necessary research tools, including genetic information and materials, these OECD Guidelines clearly support the current industry licensing practices aimed at fostering innovation in the genetic area. For instance, license agreements should always permit licensees (industry or academic) to develop and further improve the licensed genetic inventions. Demand of reach-through rights (royalties) for licensed research tools are discouraged and, if any royalties would be justified, the license agreement should then include a mechanism to set a reasonable overall royalty burden in the event of multiple royalty-bearing license required to commercialize a product. Also when licensing out to academic research, the industry will usually minimize delay of publications for patenting purposes and will be reasonable in the circumstances.

8.3 Intellectual Property Dangers

When evaluating the business or technology of another, attention to IP rights, such as patents and trade secrets, is sometimes underestimated. The technical aspects and business prospects are of course of paramount importance, but surprises about the relevant IP are not unusual and thus can deeply affect the value of the transaction or even be a deal breaker. After the closing of an agreement, IP problems can also result in costly and unpredictable litigation, or result in the sale of key products or technologies being injuncted, with a consequent revenue stop or research program halt. Therefore, it is not sufficient to simply rely upon the due diligence on technical aspects and business prospects, and afterwards in representations and warranties in the closing agreement. An independent and thorough investigation of the pertinent IP during the transaction is the first line of defense against problems further down the road.

An IP due diligence consists mainly of finding and reviewing all pertinent information of a licensor or seller in order to assess IP problems and obligations. There are two main sources: information obtained through requests to the seller or licensor and publicly available data. The starting point will be a request that the seller or licensor provides under a confidentiality agreement relevant

10) www.oecd.org/sti/biotechnology/licensing.

information on IP relating to the potential transaction. The confidentiality agreement enables the pharmaceutical company to review and assess proprietary information of a seller or licensor, and contains in general (i) restrictions on use of proprietary information received, (ii) restrictions on who can have access to proprietary information and (iii) may also foresee the return or destruction of proprietary information if no deal is reached. Although most information will come from the licensor's answers and information, there is a certain amount of due diligence that has to be conducted using outside sources. The review of publicly available material will permit verification of the information provided by the licensor.

The objective of an IP due diligence is to correlate the key IP assets (products, technology, patents) of the transaction with the market or potential market involved. The strength and limitations of IP should be particularly highlighted. The evaluation of the exclusive nature of the transaction requires an identification of all entry barriers for potential competitors and any third party IP that would affect such exclusivity. In this context it is critical to determine whether a viable alternative exists to the technology and/or products subject to the transaction. Evaluation of proprietary IP of a licensor will have to establish their strengths with respect to duration, enforceability and scope, and also clearly set out their limitations with respect to potential market entry of competitors and third-party IP.

IP dangers will have to be reflected in the deal structure, particularly for late-stage deals (costs, indemnifications, representations, etc.). Eventually, they can be a basis to stop any further negotiations. The main IP dangers to be identified during the IP due diligence and addressed in the conclusions are whether:

- Key IP of licensor/seller or of competing third party cannot be licensed or acquired.
- Exclusivity is not assured because of existing liens, joint ventures or co-owners.
- There are similar non-infringing alternatives feasible from a medical viewpoint or a commercial viewpoint.
- There is no confidence that IP will be enforceable against third parties (e.g. because of narrow protection or invalidity grounds).
- The transaction itself may cause change or loss of key IP (e.g. assignability of agreements in the event of a technology purchase).
- The transaction may impose unwanted IP obligations (duty to disclose, to keep secret, royalties, grant back, anti-trust).
- Buying an IP lawsuit (i.e. existing claims, threats, suits, or new threats created by change in competitive picture or arrival of a 'deep-pocket' pharmaceutical company).

8.4 Industry Out-Licensing Prospects

As stated previously, pharmaceutical companies are actively developing and acquiring the necessary tools for discovering and developing biologics, so most deals are in-license partnerships; in 2004, only about one-fifth were out-license agreements for the large pharmaceutical companies.[2)]

In essence, out-licensing has to provide certain benefits for the pharmaceutical industry in terms of bottom-line monetary improvements and in terms of maintaining a competitive advantage. The reasons for out-licensing can be quite diverse, such as:

- Maximizing asset value and generating new sources of royalties.
- Uploading less promising drug candidates.
- Building specialized patent clusters for cross-licensing.
- Managing and resolving threats.
- Securing technology improvements.
- Driving ethical corporate responsibility.

Given the costs of drug development in particular in the pharmaceutical field, pharmaceutical companies have to prioritize their financial resources in developing only the most promising innovations into therapeutic products. Some drug candidates appearing less promising can then be partnered or licensed out to foster development of products that otherwise would not rapidly benefit to healthcare services. For maximizing the chance of successfully developing an out-licensed drug, it is important to select an out-licensing partner that is already having some good experience in drug development, and that is financially stable to sustain the necessary lengthy and costly drug developments. For instance, over a quarter of the recent out-license partners of major pharmaceutical companies are other large companies (i.e. having more than US$ 50 million in 2004 sales).[2)]

Also, another incentive for out-licensing is to include a 'grant-back' clause whereby any improvements made by the licensee is licensed back to the licensor, generally royalty-free. In effect, this makes all licensees a further technology source. Patented research technologies of the pharmaceutical industry in fact represent a good promotional vehicle to foster additional specialized arrangements with universities or start-ups in order to develop the next technology generation and therefore maintain technological leadership.

The availability of patent protection for genes, proteins and biotechnology research tools may also be seen as an opportunity to create significant out-licensing revenues. For instance, one may use these patent rights to make money from useful research inventions without having to spend vast amounts of resources to develop drug products. The universities and start-ups have indeed developed a very aggressive patenting and licensing strategy in this area, which in some respects have been at the expense of the legal systems.

The reality for the pharmaceutical industry is different. There are always many alternatives that can be used in biologics' drug discovery and for this reason patented research tools are rarely perceived as providing a long-term competitive advantage. Out-licensing of research tools is therefore not a priority. Even those genes and proteins having the potential to be directly used in therapeutic biologics are subject to few commercial licenses, mainly through opportunistic patent freedom licenses. In fact, there are still a few biologics on the market that limit the number of potential patent freedom licenses. Also, the therapeutic potential of genes and proteins is difficult to assess, with significant investments needed, and for this reason internal R&D is preferred to out-licensing. None of this of course is to deny that genes and proteins may be valuable inventions entitled to a fair compensation.

8.5
Industry In-Licensing Practices

Technology transfers to the pharmaceutical industry are made through various types of transactions. It can be made through a patent licensing or patent purchase, but most other forms of technology transfers involve not only patent rights but also know-how relating to a specific product or technology. Therefore, R&D collaborations, manufacturing contracts and even MTAs qualify for technology transfers.

Options to license patents or mix of know-how and patents are in many instances the necessary precursor to a license of a technology or product. Companies will recognize the potential advantage of a new patented technology or product, but because of the early-stage nature of the work this may need more confirmatory evidence. Depending on the exclusivity and the period of the option, which is normally of the order of 6 months to 1 years or more, then patent costs might be a charge. The option is very much a time-dependent agreement and such limited grant will come to an abrupt end at the end of the option period.

Patent freedom licenses may be required in situations where patent threats are viewed as insurmountable. Infringement of valid patent rights can indeed result in costly and unpredictable litigation, or result in an injunction being made in the sale of key products or technologies, with a consequent revenue stop or research program halt.

If the patent covers a commercial end-product, generally royalties and milestones represent future payments. A milestone payment is a means of generating cash for the licensing organization and is more common in industries where government regulation plays an important role. There are nominal industry comparisons on royalties, but given that the occurrence of royalty stacking (through multiple patent freedom licenses) in the field of biotechnology is generally higher than in other pharmaceutical fields, the parties usually find an arrangement on royalties that does not compromise the economic break-even. Generally, partners negotiate an anti-royalty-stacking clause whereby the royalties will be lowered to a minimum if an additional royalty-bearing license is really required.

Complications arise, however, if the licensed technology is only one ingredient in a finished product. For instance, in the field of biotechnology, therapeutic antibodies have been traditionally selected through phage-display techniques,[11] but

11) Clackson, C. and Lowman, H.B. (eds) (2004) *Phage Display: A Practical Approach*, Oxford University Press, Oxford.

none of the patents covering these techniques actually cover any therapeutic antibody (i.e. any commercial product). Therefore how important is the licensed technology for the final product? On what basis is one paying and for how long? If it is acceptable to pay royalties on a product not covered by the patent claims, is it also acceptable to continue to pay royalties after the technology patent expires? Are there any limits for these 'reach-through' royalties?

If no royalties are due for a patent freedom license, then annual minimum payments are certainly made to encourage the licensee to get on and use the technology. An exclusive license would also be expected to include all future patent costs and even sometimes reimbursement of the prior patent costs. Occasionally there is also pressure from universities and start-ups, for a non-exclusive license, to include at least some element of patent costing, although this practice is arguably unfair if the licensor has then the ability to charge only one of several non-exclusive licensees. When the licensor asks that the licensee pays the patenting cost, such licensee must think carefully before agreeing to this. An unlimited obligation is comparable to handling over a blank check to the licensor, or rather to his patent attorneys (i.e. no invoice sent to the licensor will ever be queried because the licensee is paying). At the very least an upper limit must be put upon the amount to be paid.

The question on exclusive versus non-exclusive licenses can be a tricky decision and depends largely on both parties' understanding the needs of the other. Clearly, the pharmaceutical companies would like to protect their competitive position and gain exclusivity whenever possible. If the field of application is perhaps limited and there are a small number of potential licensees, then an exclusive license may be easier to negotiate. The more difficult problem arises when the technology is ground breaking, with the potential of many licenses. Under such circumstances, a university or start-up might well want to obtain several non-exclusive licensees, whilst the company would want exclusivity.

Research collaborations may get started with a simple MTA regulating the transfer of therapeutic compounds or research materials from the industry to a university or even a start-up. The pharmaceutical industry has a general practice of supplying its own compounds and research tools for research purposes, wherever justifiable, free of charge to external collaborators, contract service companies and scientists at academic institutions. However, the industry only tends to fulfill the request if it is in its own interest (e.g. if the recipient could provide useful data to promote a development or marketed compound) or to foster scientific relationships in general. The usual requirements for transfer of a material are that the industry has the right to transfer the material and that, except for pure research materials, there is some patent protection covering the materials. Furthermore, the agreement should clearly provide for the industry to maintain full ownership of therapeutic compounds and that the research result made under the MTA (in using these compounds) be licensed to the industry or at least subject to an option for a license.

The most common research collaborations may take the form of sponsored research, whereby only the university or start-up researchers will work on a research plan, or the form of joint collaborations, whereby each party contributes to a common research program. The industry funding a research program is of course aiming to protect the competitive position and therefore is asking for ownership or exclusivity whenever possible. The university or start-up has, however, other goals than simply receiving a fee for the service. For instance, in issuing a so-called exclusive license, a university will always want to retain its rights to use the technology for research. This is fundamental to any deal as the university's prime aim is to carry out educational activities, which include teaching and research. Also, the university will always want to publish the results of such research activity, but will always consider the business interests of its licenses in so doing for patents, but there are of course time limits on any requested limitations.

If the research collaboration is aimed to develop a pre-existing technology, either owned by the industry or by the university, it is inevitable that improvements to the technology are made in the period subsequent to the research term. Therefore, it is probably essential to the industry that access should be available to such improvements, whether through an automatic license or option for a license. On the same ground, the industry may also be called to license back to the university any of its improvements, but with restriction to use for academic purposes only. This 'call-back' license requirement is particularly acute in technology transfers originating from start-ups, as a mean to maintain technology leadership and therefore keep their business model intact.

The control of the patents generated out of a research program generally follows the exclusive nature of the transaction. If the industry gets an exclusive license, it is often more practical for the university that the patents are controlled by the industry (i.e. paying for the patent costs and controlling the patent prosecution, maintenance and enforcement), subject however to the rights of the university to get these patents back should the industry decide not to maintain them any longer. For many institutions and academic institutions, the financial burden of the prosecution and maintenance of patent rights, in what may be a large number of countries, is not indeed one which they can easily bear, particularly before royalty revenues start to flow in.

Research collaborations aiming to understand the basic biological mechanisms in humans, such as exploring biological pathways influencing the creation of tumors, can generate patents with a scope extending beyond the mere nature of the research work. For instance, the discovery of a new biological pathway or target can lead to the patenting of any therapeutic antibodies inhibiting such a pathway or target. This patenting activity does not even require the university to actually make (produce) any antibody. All that is legally required is a patent description containing an enabling disclosure over the invention (i.e. describing in detail how to make an antibody[12]). The availability of patent protection for genes and

12) Jaenichen, H.R., McDonell, L. and Haley, J.F. (2002) *From Clones to Claims*, Heymann Intellectual Property, Cologne, pp. 101–25.

research tools therefore constitutes a great opportunity for universities and small companies. They can use these patent rights to make money from useful inventions without having to spend vast amounts of money to develop drug products themselves. For pharmaceutical companies, on the other hand, this creates problems, since now they may have to pay to use tools that previously might have been published and freely available.

Therefore, the negotiation of a research collaboration in the field of biotechnology can be a difficult exercise. For instance, the university or start-up may insist on a high royalty component in the event any future product of the industry would be covered by patents generated during the research program. Problems arise also when the university or start-up has an inflated idea of the value of their inventions. They may well insist on getting 'reach-through' royalties on innovation relating merely to research tools.

What is a research tool? One definition is that a research tool is a composition of matter or a process that is used primarily in the research laboratory. In the context of pharmaceuticals, it is normally either a material (biological or otherwise) used in the selection or testing of a drug candidate, rather than a component of a medicament; or a process of screening or testing, rather than of manufacturing or using a drug. Examples of research tools include markers, assays, receptors, genes, transgenic animal models, etc. The inventor of a new research tool that is of general applicability (e.g. the polymerase chain reaction, green fluorescent protein, etc.) may be satisfied with compensation based on the sale of reagents or an annual fee for use. However, if the tool is more specific (e.g. a receptor associated with some disease state), they would clearly like to get a share of the final result – sales of a drug interacting with the patented receptor. Their problem is that the patent for the receptor or a screening assay using it will unlikely cover the final drug product.

None of this of course is to deny that a research tool may be a valuable invention, and that the inventor is entitled to patent protection (assuming the criteria for patentability are met) and to a fair compensation for the use of the invention by others. However, the problem is to find common grounds for a reasonable and fair compensation. It is usually not correct to say that drugs 'would not have been discovered' without the patented research tool. There may have been alternatives which could have been used. In the United States, the National Institutes of Health (NIH), which provides a substantial amount of the funding for academic research, has taken the clear position that imposition of downstream royalties are not appropriate for recipients of NIH funding.[13] The NIH hopes that other not-for-profit and for-profit organizations will adopt similar policies and refrain from seeking unreasonable restrictions or conditions when sharing materials. In the same perspective, the OECD Guidelines on licensing genetic materials discourage reach-through royalty practices.

13) *Federal Register*, **64**, 248, 23 December 1999.

Late-stage technology transfers concern more the transformation of a third party's product or technology into a commercial success and less the research for drug discoveries or technology improvements. These technology transfers can have the forms of in-licensing (including purchase) materials and proprietary information relating to a marketed drug product or a drug in clinical development, up to a manufacturing contract whereby production technologies are exchanged. Given the commercial end nature of these transactions, the financial aspects and the roles of each party in the future regulatory proceedings, supply and commercial sales are critical aspects to be negotiated.

Although the in-licensing contracts all contain elements in common with, for example, patent freedom licenses and research collaborations (see above), there is no such 'standard license' – there will always be unique clauses in any such agreements. An important trend is the fewer in-license deals on clinical drugs in which the licensee has the full global marketing rights. This implies finding a proper market power for each partner and therefore a careful negotiation of the roles is required within the framework of competition laws. Some unique clauses will be drawn to address IP dangers that have been identified during the due diligence (see above). Other clauses will carefully organize the drug development strategy, which is critical to achieve drug approval, knowing that the major reason for contract termination of license deals for clinical drugs is in fact a scientific reason, more than a legal or commercial reason.

About one-third of the new molecules in clinical phase III are now in-licensed, and the pharmaceutical companies have to continuously access a vast reservoir of technologies and drug products in clinics. In this equation, the start-up companies play a central role for the discovery of novel therapies, with an almost infinite appetite for capital. The industry sees a responsibility in stimulating this innovation and therefore makes early investments to develop such companies. This complementary approach to internal R&D and licensing activities has been pursued, for more than a decade, through internal venture units of companies that are clearly separate from R&D. This funding activity has the goal to early identify the most promising novel therapies or technology opportunities, but also to generate money that is continuously reinvested in new commitments for the discovery of novel therapies. This practice has proven in many instances, as the start-ups grow, to foster collaborations and business deals with the sponsor industry.

8.6 Challenges Related to Biotechnology

8.6.1 Experimental-Use Exemptions

In the *Merck v. Integra* case, before it was overturned by the Supreme Court, the damages awarded by the jury for a little research infringement were US$ 15 million. The Supreme Court held that Merck's actions were subject to the Safe Harbor, hence no damages were payable. However, the underlying message remains intact – a research infringement can result in damages comparable to the kind of payments ordinarily made in licensing transactions, which can be quite substantial. The Supreme Court left new sets of unanswered questions, in particular as to the limits of the USC §271(e)1. Consequently, the case is closely watched by companies that make their income based on patented inventions that are used in the development of data for Food and Drug Administration submission. For example, does the statute exclude research infringement using patented technologies like instruments and techniques for DNA and proteins? From the perspective of the pharmaceutical industry, this uncertainty has at least the positive effect to create a more balanced negotiation power, hence making it easier to negotiate reasonable and rationale licensing terms for biopharmaceutical research tools.

8.6.2 How Essential are Patented Genes?

A 2002 report by the OECD on the effects of patents in blocking research and commercial developments, stated that:

> ... the few examples used to illustrate theoretical economic and legal concerns related to the potential for the over-fragmentation of patent rights, blocking patents, uncertainty due to dependency and abusive monopoly positions appear anecdotal and are not supported by existing economic studies.

These conclusions force to question the assumption made by academics that there is a blocking or restricted access to technologies in the biopharmaceutical field, just looking at the growing number of biotechnology patents.[14]

In this context, the more recent OECD Guidelines for the Licensing of Genetic Invention (2006) provide a new interesting framework encouraging the licensing of genetic research tools at reasonable cost, by stating, for example, that 'rights holders should broadly license genetic inventions for research and investigation purposes' and 'license agreements should avoid reach-through rights'. From the perspective of the pharmaceutical industry, the Merck case combined with these guidelines create good conditions for the swift licensing of genetic research tools and therefore the promotion of research-based innovation. This is critical for the endeavor to address unmet medical needs and develop novel therapies.

The OECD approach is, however, a difficult one for smaller biotech companies, or even some universities, that strive to generate revenues from patented research tools. Basically their business model consists of licensing genetic research tools, sometimes exclusively, with expensive up-fronts and reach-through royalties to

14) http://www.oecd.org/sti/ipr-statistics.

any commercial product (although their IP does not cover it). The controversial article in *Nature*[15] is an example of adverse consequences. It describes the case of Jeanne Loring, an embryologist at the Burnham Institute in La Jolla, California, claiming that her start-up firm collapsed when it could not get access to embryonic stem cells at a reasonable price from the Wisconsin Alumni Research Foundation, that owns seminal patents in embryonic stem cell research. As a result, assumptions were drawn by academics and political pressure groups that different collaboration models should now be developed to encourage research (e.g. by combining a patent pool, an open-source model of IP development and/or a shared prize system).[16]

There are of course some licensing problem areas, often in relation to the delays and transaction costs involved in licensing, but there is no mounting evidence that the licensing system is not working. In fact, in many cases, there seems to be an 'informal research exemption' (i.e. many research tool patents are being ignored or not enforced, particularly in the context of academic research). Many patents are filed for defensive reasons and therefore not enforced. Many patent holders encourage use by academics to make the tool widely known; however, many simply do not have the resources to enforce patents if this means investigating what goes on in a laboratory behind closed doors.

One difficulty for the pharmaceutical industry may, however, reside in the OECD Guidelines to provide broad commercial and non-exclusive license to foundational genetic inventions. Although the large pharmaceutical firms are certainly open to discuss new collaboration models, it is important to keep in mind that they need the right financial incentive and the respect of IP for using their manufacturing, mass-screening and clinical trial management skills. Therefore, exclusive arrangements may be often required in order to recoup the necessary investments. The same is also true in the diagnostic area, understanding, however, that the licensing policy of Myriad Inc. is not any good model (i.e. they demanded high prices for testing under the patented technologies and prevented other laboratories from carrying out any diagnostic testing).[17] This licensing practice of Myriad raised inflamed ethical questions and led to a range of European government interventions until the Myriad European patent was finally revoked in 2004.

FP6 is the main instrument of the European Commission for funding health research and ultimately improve the quality of life.

However, a key hurdle for the industry is to fully exclude from the automatic licensing to other participants any future proprietary (therapeutic) product. There is indeed a requirement to license to other participants any know-how relating to the research program – that includes any future know-how developed within or outside the research program! If the FP6 research program is aimed to discover

15) Wadman, M. (2005) Licensing fees slow advance of stem cells. *Nature*, **435**, 272–3.

16) Goozner, M. (October 2005) A new paradigm for managing intellectual property at the California Institute for regenerative medicine, California State Senate.

17) http://www.wipo.int/wipo_magazine/en/2006/04/article_0003.html.

new drugs, in an area where the pharma industry is already active, or even intends to be active, an inevitable overlap will arise and the risk is then to have disputed exclusivity or royalties over proprietary products. There are of course legal mechanisms to manage such risk, such as by excluding specific existing drug molecules, imposing the respect of confidentiality, not authorizing sublicense rights, including grant back licenses over improvements and limiting the commercial access rights to a maximum of 2 years. However, in practice these clauses are difficult to negotiate with multiple participants and the exclusions cannot cover the unknown (i.e. future drug discoveries made by the industry during, and related to, the research program).

Not to deny the real promotion of research brought by FP6, the large pharmaceutical companies tend either to walk away from FP6 or limit its participation to basic research work that is away from any significant commercial outcome.

One can only encourage more biologics research projects and multiplication of commercial projects, with the key participation of large pharmaceutical companies in using their manufacturing, mass-screening and clinical trial management skills. It is positive to read that the FP7 for the years 2007–2013, which is still in elaboration, is now recognizing the need to attract increased private investments. The European Parliament legislative resolution of June 2006 on FP7[18] also provides clear support to more industry participation to FP7 and commercial exploitation, when stating that 'the participation of the private sector and the commercial exploitation of scientific and technical results should be encouraged but a balance between IP rights and dissemination of knowledge should be found' or when stating that 'the participation of the business sector and the commercial exploitation of scientific knowledge and technical skills are important factors in ensuring that the seventh Framework Program does in fact make a contribution to increased growth and the creation of jobs'.

8.7 Biodiversity Convention

The Eighth Conference of the Parties to the Convention on Biological Diversity decided in May 2006 to establish a group of technical experts to explore and elaborate possible options on creating and using recognized a certificate of origin ('Certificate') as one possible means of achieving the objectives of the Convention in relation to access and benefit sharing ('ABS'). The impetus to create the regime came in part from the recognition that many developing countries were resource-constrained in their ability to police ABS effectively and the associated concern that without an international element, policing would be impossible because users and providers are often based in different national jurisdictions. These original ideas have been progressively overshadowed politically by concerns about

18) Text of the European Parliament, P6_TA-PROV(2006)02.

the need to prevent 'rampant bio-piracy'. However such an issue is not as prevalent as some believe. For instance, Biodiversity-relevant resources that have a one-to-one relationship with a final product are rare. Where they exist, they are normally sold as 'natural' products and, in complete contradiction to the idea of bio-piracy, their origin is an intrinsic part of the way they are marketed.

Therefore, it becomes important to critically evaluate whether and how Certification, as an element of any international regime, can really address ABS. It should indeed be practicable, transparent, efficient and, above all, should avoid arbitrary treatment, consistent with the provisions of the Biodiversity Convention. Given the complexity of creating a Certificate regimen, the industry is overall rather in favor of a contract-based approach in a transparent manner to address the same objectives. For instance, best-licensing practices could be elaborated and then incorporated in the ABS laws of countries. Thereafter, any contracts should follow these standards and could be reviewed by competent authorities. Such best-licensing practices could include auditing discoveries made out of the genetic of information, royalties obligation for significant discoveries, exchange of discovery materials, training of local scientists, etc.

8.8
Industry Motivation Driving R&D Investments

Understanding the legal regimes and practices for technology transfers does not enlighten the principal motivations driving the industry R&D investments. What are the critical factors that a company has to take into account?

Mr Daniel Vasella, CEO of Novartis, attributed the R&D lead that the United States has over Europe to the much larger federal budget and a venture capitalist system that is extremely effective at investing money in pharmaceutical technologies.[19] Also, traditional US academia has the freedom to create companies and be active in enterprise as well as academic research. It is slowly going this way in Europe, but for decades this kind of activity was not really allowed, or was seen as inappropriate behavior. The United States also has a more open and progressive 'can-do' culture. Education plays a role in the equation and the greater concentration of good universities in the United States has helped the United States achieve nearly double the PhD rate, per head of the population, than is seen in Europe. In terms of the number of publications and citations, the United States is also far ahead.

Far from a 'can-do' culture, Europe, has a much more cautious approach. It is more conservative, with the anti-genetically modified organisms, anti-genetics, anti-patents movements, for example. This influences policy makers and politicians. The United Kingdom faces a particular problem with animal rights extremists – one which is not experienced elsewhere in the same way. Companies that cannot afford the sort of security they need to protect their employees and

19) 2003 interview to the *Financial Times.*

facilities switch their investment somewhere else. There are simply many more balancing acts that you have to do in Europe.

The real value creation for pharmaceutical companies is in their IP. Developing countries like India or China will only get scientific investments by industry if they have adequate protection for IP. India may become one of the 10 most important pharmaceutical markets in the near future. The problem is that few want to go into India to do discovery. Why? Because there is limited patent protection and no data exclusivity.

All these factors have given the United States the essential ingredients for a successful biotech environment. Other countries should strongly encourage protection for IP to drive R&D investments locally.

Threatening of compulsory licensing to resolve globally the issue of access to medicines, followed by negotiation with pharmaceutical companies, is clearly a deterrent for R&D investments.

In fact, the impact of patent protection on patients' access to treatment in developing countries is exaggerated: although more than 300 out of 319 of the drugs on the World Health Organization (WHO)'s model list of essential drugs are either available off-patent or not patented, over a third of the world's population still has no access to essential drugs. As essential drugs for the management of HIV/AIDS and drug-resistant forms of tuberculosis and malaria are patent-protected, and therefore the exception to the rule, innovative and unorthodox solutions must be found to improve poor people's access to these treatments.

Compulsory licensing should rather be limited to real emergency situations, particularly to not undermine benefits of the patent system. The non-respect to patent protection must be understood indeed to inevitably lead to a reduction of investment, discourage research or encourage researchers to adopt a policy of secrecy, limiting the access to technology and novel therapies. In reality many people in developing countries are now taking greater advantage of the patent system to establish solid businesses in their country and abroad. The example of India, with flourishing local pharmaceutical companies (not only generics) filing patents, is a symbol of a positive evolution that may create a more balanced debate on compulsory licensing.

In an attempt to address the issue of access to medicines, there are also a growing number of pharmaceutical companies, backed by rising government and philanthropic support, seeking to develop drugs and vaccines for diseases of the developing world where ability to pay is extremely limited. For instance, Novartis has been the first to launch an internationally approved and effective therapy for malaria that uses a derivative of artemisinin – a Chinese traditional medicine. However, the company was then drawn into a dispute over production levels, pointing to the WHO of exaggerated forecasts of 120 million treatment courses a year and finally being accused by medical charities of failing to produce enough. Thereafter, Novartis was forced to subsidize production of Coartem, in an effort to boost demand from poor countries in Africa and Asia, where malaria is endemic.

This open questions about what can be incentives for companies to systematically invest in developing drugs and vaccines for diseases of the developing world.

Drugmakers are so often vilified by patients and politicians alike, because of what they see as unreasonably high drug costs.

The good new is that the industry is clearly committed to behave impeccably, has no reasons to be defensive about its actions and is doing more positive things for the society. Novartis has a tremendous track record in taking the high ground and donated US$ 570 million in 2004 through various corporate citizenship programs.

Novartis is committed to helping improve patients' access to its treatments for diseases of poverty. Novartis has signed two Memorandums of Understanding with the WHO – one to provide free treatment for all leprosy patients in the world until the disease has been eliminated from every country and the other to provide Coartem, its oral fixed-combination anti-malarial product, at cost.

Novartis is also committed to supporting *pro bono* research on diseases of poverty. It has establishing a research center in Singapore with a focus on developing new preventive and effective treatments for tuberculosis and dengue fever. The diseases affect 2 billion and 50 million people, respectively, mainly in developing countries. The Novartis Institute for Tropical Diseases is a result of an agreement between Novartis and the Singapore Economic Development Board and involves an investment of US$ 122 million.

9
A Quarter Century of Technology Transfer in US Universities and Research Institutions

Lita Nelsen

9.1
Introduction

Technology transfer in the United States is nearing maturity – after a prolonged adolescence. In this chapter, I will briefly cover the history of technology transfer from universities and research institutions over the past quarter century and its impact on the US economy, and will then cover some of the issues arising from the success of university technology transfer. For simplicity, the term 'universities' in this chapter is used broadly to cover academic institutions, medical research institutes and hospitals, and other non-profit research institutions engaging in fundamental research. Although the degree-granting academic institutions dominate this research, the other institutions are important players, sometimes with slightly different missions. The governing legislation and the technology transfer practices are very similar for all.

Although a few universities had engaged in patenting and licensing of the inventions arising from their research as far back as the 1930s, with some acceleration in the 1960s and 1970s, widespread participation in this activity began in 1980 with the passage and implementation of the Bayh–Dole Act. This groundbreaking legislation affected patents from universities arising from research funded by the US Government. Since over 90% of the basic research in such institutions is from competitive grants from the US Government, this new law changed the landscape. In the following quarter century, technology transfer changed from an occasional activity into a significant thrust by hundreds of institutions; from a few hundred patents filed per year to more than 5000 US patents issued in 2005 (and thousands of foreign counterparts); and from a few dozen licenses to over 4000 licenses granted in 2005 alone.

Technology transfer is now considered to be a major contributor to the continuing growth of the technology economy in the United States, and a particularly important source of the most innovative new products in medicine, materials and information technology.

Technology Transfer in Biotechnology. A Global Perspective.
Edited by Prabuddha Ganguli, Rita Khanna, and Ben Prickril

ISBN: 978-3-527-31645-8

9.2 Enabling Legislation

The Bayh–Dole Act (Public Law 96-517, codified as USC §§201–211) allowed research institutions to own the patents arising from their federally funded research programs. It allowed the institutions full discretion in granting licenses – although it took several years and some changes in regulations, before they could grant exclusive licenses for the life of the patent. It also allowed the institutions to receive royalties for these licenses, provided that a portion (unspecified) was shared with the inventors of the patents and that the remainder was used solely for education or research.

The Act also required that the US Government be given a royalty-free non-exclusive license for government purposes (only) to any patent filed and required that any licenses granted for commercial purposes require 'diligent development' on the part of the licensees.

9.2.1 Public Purpose of Bayh–Dole

The new law was passed with the purpose of stimulating the US economy. In the late 1970s (just prior to passage of the Act), the United States was concerned about its economic competitiveness in the world. Maintaining its competitive standing would depend on US industries' adopting new technology rapidly, as the world transitioned into what is now called 'the knowledge economy'. However, although the country led the world in the quality and volume of its fundamental research (funded in significant part by the US Government), industry was slow in adopting the results of this research.

Advocates for small businesses (who also received significant federal funding for research through the Small Business Innovation Research Program) argued that if they, rather than the federal government, owned the patents arising from their federally funded research projects, they would have a much greater incentive both to participate in the government's research programs, and to invest in moving the research into product development and manufacturing.

Universities and research institutions argued on their own behalf that patents could also be important in moving the results of research from their laboratories into the economy – and urged the inclusion of non-profit research institutions in the proposed legislation. The premise here was founded on the fact that university research results were 'embryonic': they were 'laboratory stage' and would require substantial investment (millions to tens of millions of dollars and a number of years) to bring them to market; and this investment was highly risky, since neither the commercial practicality of manufacture nor the potential acceptance of the market was known. A company had very little incentive to begin development of a university invention, given the magnitude of both the investment and the risk if it would be threatened by competition as soon as it showed the practicality of the innovation.

The universities argued that they could use patents on their inventions to solve this dilemma, by protecting the 'first mover' innovating company from its competition. The equation was fairly simple: if a competent company agreed to take the risk and make the investment to try to develop the product, the university would grant it an exclusive license. If the product succeeded, the university's patent and exclusive license would protect the company from competition coming in. The period of time for which the university would grant exclusivity would be commensurate with the investment (in both time and money) at risk to bring the product to market.

9.2.2 Returns and Economic Development

In the quarter century since the passage of the Bayh–Dole Act, technology transfer from universities has blossomed, with more than 200 universities and research institutions reporting significant activity. In fiscal year (FY) 2004, for example, the Association of University Technology Managers (AUTM)[1] reported 3680 new US patents issued (and thousands of foreign counterparts), over 4000 licenses, of which 658 were to new start-up companies, formed specifically to commercialize a university-licensed technology.

The total royalty revenue received by these universities in FY 2004 was US$ 1.4 billion (which includes both cash payments and cash-in of equity taken through licenses to start-up companies). Although this is a substantial sum, it should be remembered that this is from the *entire* US university and research institute community, whose combined research budget is over US$ 41 billion. Thus, the technology transfer revenue constitutes only 3.4% of the research volume that generated it. A closer examination of the data shows that the majority of the revenue is attributable to a relatively small number of 'blockbuster' patents, with the revenue being very unevenly distributed among universities. A few universities contribute a substantial amount to their universities' operations during the lifetime of a blockbuster patent, while the majority of technology transfer offices do little more than break even financially. This is after 25 years of operation.

The financial findings would indicate that technology transfer should not be looked upon primarily as a potential source of significant funding for a university, since few universities can hope to get lucky with a blockbuster patents. Instead, technology transfer needs to be seen as a long-term investment in societal impact, through innovative new products, technology that increases economic competitiveness, and creation of new industries and new jobs.

It has been estimated also that approximately 400–500 000 jobs in the United States are the direct result of university licensing, with employees working either

1) Association of University Technology Managers (2004) *FY 2004 Licensing Survey*, AUTM, Deerfield, IL. http://www.autm.net/surveys/dsp.surveyDetail.cfm?pid=28.

in developing or manufacturing and distributing products based on technologies licensed from universities. Of course, each of these 'direct' jobs results in economic flow from salaries that result in two to three other jobs being created.

9.2.3
Start-Up Companies and Entrepreneurship

In the 1990s, technology transfer offices in universities began to emphasize licensing to start-up companies based on their technology. This trend accelerated in the mid-to-late 1990s and is still continuing. Although the majority of university licenses are still to existing companies, start-ups are becoming increasingly important – particularly in local economic development.

This emphasis on start-ups resulted from a number of factors, including among others: (i) an increasingly short-term earnings focus of large companies, such that they were reluctant to license and invest in new technologies that would take many years to develop; (ii) a very large increase in pension capital and other capital available for venture investing funds; (iii) successful start-ups in a few geographical regions that showed the economic impact of entrepreneurial companies – and encouraged some state legislatures to fund activities to promote entrepreneurship; and, perhaps most importantly, (iv) role models: faculty who started companies that prospered – leading to significant enrichment of the founders and sometimes also of their universities.

The emphasis on university start-ups has had a major impact on a number of industries. The majority of biotechnology companies, for example, can trace their origins directly to licensing of university technology. The great biotechnology clusters in the United States (such as Boston/Cambridge, San Francisco and San Diego) all formed around research institutions with major biology faculties – and significant federal funding of their research. A good many information technology and software companies (Google! out of Stanford University being a notable case) also trace their origins to university-based research.

These start-ups provide many advantages and opportunities:

- A start-up company, coupled with the consulting privileges available to faculty under most university policies, provides an attractive mechanism by which a faculty member can continue with his or her fundamental research at the university while participating actively in translating that research into practical utility. This mechanism also benefits both the company and the university: the company gains the vision and in-depth knowledge of the science and technology that the faculty member brings, while the university retains the teaching and research skills of the faculty member. All benefit, both financially and through their impact on the society, if the company succeeds.
- In contrast to a license agreement to an established company, a license to a start-up company virtually assures that development of the licensed technology will be a high priority for the licensee – at least for the first few years of the company.

- New entrepreneurial companies can have a significant impact on local economic development, creating new high paying jobs. If a cluster of companies in the same industry forms in the region, further development of new companies in that industry becomes easier – because of a concentration of investors, management talent and suppliers in the region.
- Entrepreneurship breeds entrepreneurship: students and faculty become alert to the opportunities through watching others succeed; new courses in entrepreneurial business practices are developed in the business schools; venture capital funds may move into the region to become close to the source of new companies; and an entrepreneurial ferment in the region develops

However, start-ups can also have disadvantages, which should not be discounted:

- Building a new company requires significant resources; if an existing company with technical and financial capability combined with marketing power is willing to invest in developing a new technology, it may very well have a higher probability of succeeding and growing the market for the new technology.
- Start-up companies are fragile. They may have difficulty raising sufficient capital to get the product and the market developed. They may fail because of bad management – or even bad timing (developing a product for which the market is not yet ready or in which the industry to which it hopes to sell products is collapsing).
- If the university is counting on significant economic return from its equity holdings in a company, its hopes may be dashed: if the company requires more rounds of funding than anticipated and/or if the company needs more financing at a time when it is not doing well, then the equity shares may be diluted to the point where the university holdings are worth very little. This is not an uncommon occurrence, and is one reason why many universities look for royalties on products in addition to equity holdings from their start-up companies.
- Finally, conflicts of interest inevitably arise with start-up companies. Universities need to set clear policies to assure that faculty members' commitments to the university take precedence over their commitments to their companies. The university itself must draw 'bright lines' between its academic work and its relationships with its start-up companies, assuring itself and the public that the academic mission takes priority. Managing the conflicts of interest takes considerable administrative talent and time.

9.3 Lessons Learned in a Quarter Century of Technology Transfer

There is little doubt of the ultimate potential of university technology transfer programs to accelerate the adoption of new technologies in industry, to enhance entrepreneurship, to create new medicines and other products for the public, and to create jobs and add to prosperity through economic development. The clustering of high-technology and biotechnology companies around major universities has been well described, and the creation of hundreds of thousands of jobs directly related to university licenses and start-ups well documented by the AUTM and others.

Within universities, robust technology transfer programs have many important positive effects, quite separate from monetary return. These include, among others:

- Productive interaction with the industrial community, with ideas shuttling back and forth between the academy and the private sector, often increasing the quality of research.
- Increased industrial support of university research.
- Increased willingness of central and local governments to support research at the university in the cause of economic development.
- Student exposure to industrial ways of thinking and to identifying commercial opportunities from research.
- Student exposure and training in entrepreneurship, influencing their future career aspirations and ultimate impact on the country's economy.
- Financial support from grateful alumni and other entrepreneurs who have grown wealthy from companies started from university research.

For the surrounding region, such programs can also have a major impact on the economy – not only from direct entrepreneurial spinout companies from the university. The entrepreneurial ferment and capability resulting from university spinouts in turn leads to the formation of many other new companies and often to the moving in of larger companies which seek to take advantage of relationships with the entrepreneurial companies and the skilled employee base in the region.

9.3.1 Expectations in Setting Up a Program

Despite the promises that can be expected from successful technology transfer programs, it has been and continues to be a rocky road when communities and their universities try to start new technology transfer programs or accelerate existing ones. Unrealistic expectations are a major cause of failure and frustration in building these programs around the world. Universities not only expect such programs to bring in industrial sponsorship of research programs, but often expect the royalty income or equity returns from entrepreneurial start-ups to provide substantial support for the university as a whole.

Government expectations are also often equally unrealistic. Too often governments, both local and national, expect that just a few years of financial support for technology transfer, coupled with pressures on the universities to produce measurable impacts, will almost instantly result in thriving clusters of biotechnology, software, or telecom companies akin to those in Boston, Silicon Valley or San Diego.

Almost a quarter century of experience in US technology transfer gives a more realistic picture.

9.3.1.1 Licensing Income

As discussed earlier, even before subtracting expenses for patenting and staff costs, technology licensing and start-up equity income averages *less than 4%* of the research base of US universities. There are only a handful of 'blockbuster patents' and fewer than a handful of highly profitable start-up companies per year in the entire United States.

Thus, universities should be looking primarily to the other benefits of technology transfer from their programs, with the occasional large royalty return – if it happens – seen as a bonus of the program.

9.3.1.2 Building a Program Takes Time and Money

Studies have shown that it can take a technology transfer program 8–10 years or more to reach profitability, but the majority of programs become profitable if the effort is sustained to build them.[2)]

Universities with smaller research bases have a more difficult time if 'break-even' is measured only by royalty income. The smaller amount of research means that fewer inventions arise, lowering the statistical probability that a blockbuster invention will occur. Fewer opportunities for licensing also mean that the technology transfer staff gains less experience per year and therefore learns the craft of technology transfer more slowly. Thus, small technology transfer programs may have to be financially sustained for a long period of time, with the revenue shortfall justified by the non-royalty contributions to the university and the community.

Finally, it should be noted that new technology transfer programs are too often starved – both for money to file patents and for staff – with the university expecting that the program will somehow bootstrap itself into profitability and expansion. More often, unfortunately, an anorexic program only very slowly climbs the learning curve – and reaches profitability – and has a much lower impact on the university and the community along the way.

Thus, the university needs to have a well thought out, long-term financial plan for building its technology transfer office, based on the expected benefits – both financial and especially non-financial – and what it can afford over the decade or so it takes to build a program to maturity.

9.3.1.3 Culture Change

Founding a successful technology transfer program means changing a culture. Researchers must become aware of the utility and rewards of identifying potentially commercializable inventions from their research and cooperating with

2) Brandt, K.D., Stevenson, E.J., Anderson, J.B., Ives, C.L., Pratt, M.J. and Stevens, A.J. (2005) Do most academic institutions lose money on technology transfer?, Poster session at the *AUTM Annual Meeting*, Kansas City, MO. http://sites.kauffman.org/pdf/tt/Stevens_Ashley.pdf.

industry to transfer the technology. This will be a new way of thinking for most and some will feel that the process will threaten the very purpose of the university.

This culture change must start from above. The upper administration needs to clearly delineate the purpose and potential benefits of a technology transfer program – not only to the individual and the university, but to the community at large. To the extent that this is true, the administration of the university can allay mistrust by making it clear that technology transfer will not be allowed to distort traditional academic principles of investigator-initiated fundamental research and open dissemination of information within the university, and uncensored publication.

9.3.1.4 Defining the Mission

Defining the mission of a technology transfer program and its priorities must be the role of upper administration in the university, along with the faculty. Is it primarily to produce licensing income? Or industrial support of research? Or is the mission primarily to get technology developed for the public? Or is it primarily to generate start-ups and regional economic development? Or what?

There are inevitably trade-offs among these potential primary missions. Unless priorities are explicitly set, the practices of the technology transfer office may well diverge in time from the best interests of the university. Yet, even in the United States, with a quarter century of experience in university technology transfer, it is rare that discussions on mission and priorities are held between university management and the technology transfer office.

9.3.1.5 Setting the Ground Rules: Policies and Practices

The technology transfer office and the constituencies with which it deals on a daily basis – researchers, companies and investors – must know the ground rules before they begin. It is harmful to the growth and learning process of the office if each new invention or license-in-negotiation must go through a committee to decide what is allowed. Such issues as ownership of the intellectual property (IP); rights, duties and obligations of the faculty in technology transfer; sharing of revenue and equity with inventors; use of university facilities by companies; and related issues should be defined in clear policy guides as early as possible in the evolution of a technology transfer process.

A new office can find many guides from experienced universities to help write their own – but only the administration and faculty of the university itself can decide what the ground rules for a particular university should be.

9.3.1.6 Conflicts of Interest

Technology transfer inevitably brings conflicts of interest. The challenge is in managing them.

For the university itself, for example, there may be conflicts between maximizing financial return and preserving openness of publication; in deciding whether and when to allow use of university facilities and staff (or even students) in helping start-up companies; or in relaxing policies on ownership of IP or indirect cost rates in order to bring in more research support from industry. A big conflict of interest arises when university administration is called upon to make an exception to a long-standing policy in order to bring in a big program; the exception itself may be only marginally harmful to the university, but the willingness to make an exception 'for enough money' or 'for a very senior person' inevitably causes problems later.

For the faculty members, conflicts of interest may involve, for example, conflicts of time (often called 'conflict of commitment') between time spent in teaching and university research and time spent with the start-up company; there may be temptations to withhold dissemination of research data from university research because it could be useful to the company if held secret, or, on the other hand, harmful to the company if published. Use of students on company projects presents another potential conflict of interest, as does allowing the company to use university equipment. A conflict of interest arises when a researcher decides whether their new patent belongs to the university, themselves or their spinout company.

(Although not discussed further here, even a national government can find itself with a conflict of interest: does it want to support 'basic' research in its university, keeping its scientific community at a world standard and helping discover new 'frontier' technology for the coming decades, or should it shift its support to 'practical' research that will more likely result sooner in technology transfer, new start-up companies and regional economic development.)

For universities and their faculty members, well thought-out written policies consistently applied can ameliorate many of the conflicts of interest. However, there are inevitably gray areas or appeals for exceptions – which grow with time and maturing of the technology transfer program. The university needs to define a clear chain of command for ruling on most of these issues, with only the rare exception going to oversight committees; otherwise the process bogs down while waiting for committees to be assembled and convened. Twenty five years of experience tells us that exceptions to policy should be granted very, very rarely; it is difficult in a university to make an exception for one faculty member without soon being called upon to make a similar exception for the next one – and policies soon erode to meaninglessness.

9.3.1.7 **Talent**

Technology transfer officers need an unusual combination of qualifications:

- An understanding (though not necessarily as a practitioner) of research at the state of the art – often over a fairly broad range of technologies in a multidisciplinary university. Usually this requires a solid background in science or engineering
- An understanding of the 'language of industry' – how technology is developed into products, markets, accounting and finance principles, and how decisions are made.
- At least a minimal understanding of venture capital, start-up formation and small company operation.
- More than a passing familiarity with patent law.

- An understanding and a deep sympathy with academic ways of operating, academic principles, and the career development paths and aspirations of students and professors
- Outstanding written and verbal communications skills in both formal and informal situations
- Good negotiation skills – or the innate talent, intelligence, emotional control and 'people skills' to learn how to negotiate
- Ability to deal with multiple constituencies with conflicting objectives, most of whom one has no authority over
- Ability to deal with highly ambiguous, confusing situations
- Both the drive and creativity to solve complex multidimensional problems and arrive at win/win solutions
- A drive to completion – to getting the job done.
- Very high personal integrity and the wisdom not to get into situations that get 'close to the line' on ethics – no matter how profitable to the university, the faculty member or the licensing person. A university's reputation is priceless; it cannot be endangered by unethical behavior – or naiveté.
- The willingness to work at a university salary because of the inherent satisfaction of the technology transfer job – great technology, complex and always interesting issues, the satisfaction of seeing new companies form and new technologies reach the market – and the opportunity to contribute to the university, its students and the community at large.

People with these qualifications are not easy to find. However, one should not underestimate the need for talent at a very high level. Technology transfer directors speak of the profession as a 'talent-based business – some can do it, and some cannot'. Those who can do it well perform many-fold better than those who cannot – and bring much better relationships with the researchers and the business community over time, thereby building the office's effectiveness.

In choosing staff, some formal qualifications in technology and business are a *sine qua non* – and can be checked on a resume. Whether the technical background is at a bachelor's or PhD level is relatively unimportant, provided that the person is very bright, and can understand how research is done and universities operate. Unfortunately, though, the creativity, interpersonal skills, ability to deal with ambiguity and drive to completion that are required for the job often cannot be determined until the candidate has taken the job.

9.4 Conclusions

The technology transfer process in the United States – and the learning derived from its growth – is far from complete. New challenges are constantly arising. A particular challenge is in technology fields that require many years of product development before reaching market. Both established companies and venture investors are demanding a later stage of development of technology before they will invest in university technology – and the universities are looking for new mechanisms to advance their technologies to 'investment stage'. The crisis is evident in drug development, where the gap between university discoveries of the basic mechanisms of disease and a proven drug is very large. A number of experiments are being tried to bridge the 'development gap', from university investment funds to philanthropic investments (and a few state government initiatives), but the experiences are new and the practicality of these models is not yet proven.

Many of the lessons learned from the US experience will be applicable to other countries, but some will not be, because of local educational models, political and economic structures, etc. More generally, however, we can conclude that over the past 25 years, the technology transfer experience has been a very positive and rewarding experience in the United States – for the university, the researchers, the students, the business community, the public that benefits from the medicines, products and jobs that result, and the professionals helping to make the process happen. It is, however, a complex process requiring sustained dedication at every level.

10 Technology Transfer Issues in Biotechnology: The Future of Global Health Networks

Gerald T. Keusch and Ashley J. Stevens

10.1 Introduction

Global health emerged as a major worldwide concern in the years approaching the turn of the millennium, for it had become all too apparent that the twentieth century, with all of its advances in life sciences and medicine, and improvements in health, had only widened the disparities between the rich and poor countries, and between the rich and poor within countries. The concern was heightened by two additional factors: (i) the recognition that health is an important determinant of economic productivity and growth, social development, and ultimately political stability, and (ii) because health and access to healthcare were becoming more widely accepted as fundamental human rights, requiring nations to make a social commitment to provide healthcare services and improve the health status for all of their citizens. The Alma Ata Declaration of 1978 was the political expression of this dream and provided a target, health for all by the year 2000, however much of a stretch that goal was. In fact, it did not succeed and by a number of criteria, many fueled by the emergence of HIV and AIDS, health disparities grew instead of narrowed. Realizing this, and as the year 2000 approached, a set of Millennium Development Goals (MDGs) were developed to focus on poverty reduction:

- Goal 1: Eradicate extreme poverty and hunger.
- Goal 2: Achieve universal primary education.
- Goal 3: Promote gender equality and empower women.
- Goal 4: Reduce child mortality.
- Goal 5: Improve maternal health.
- Goal 6: Combat HIV/AIDS, malaria and other diseases.
- Goal 7: Ensure environmental sustainability.
- Goal 8: Develop a global partnership for development.

Although representing as much of a stretch as the goals of Alma Ata, the MDGs are at the center of efforts to improve health status and reduce health disparities

Technology Transfer in Biotechnology. A Global Perspective.
Edited by Prabuddha Ganguli, Rita Khanna, and Ben Prickril

ISBN: 978-3-527-31645-8

around the world, for at least six of the eight MDGs are directly or indirectly related to improving health or dependent on improved health.

In the intervening years since Alma Ata, television has made it possible to show the world as it is, with all its blemishes, civil wars, genocides, violence, natural disasters, famines, disease, and disastrous political leaders and worse governments. The media have also put a human face on the deadly statistics of the ongoing global pandemic of HIV infection and AIDS, aided by the many rock stars, actors and sports figures who either have been infected themselves or are otherwise moved to help to mobilize the needed resources for people in developing countries where access to life-saving anti-retroviral drugs has been limited by affordability, availability, and a lack of political will and commitment to equity in health.

HIV and the AIDS pandemic are only a part of the health gap, however important a part. There is far too little attention or resources given to other infectious diseases, with the possible exception of malaria and tuberculosis, one reason for there being a 'Drugs for Neglected Diseases Initiative', or to the pandemic of cardiovascular disease, stroke, diabetes and obesity, or to unipolar depression and other mental health concerns, or to road traffic accidents and other causes of trauma including wars and land mines, civil disputes, and violence against women and children.

The primary issue in this chapter, however, is to explore how we can mobilize new research, appropriate use of intellectual property (IP) and scientific advances to serve the benefit of all in need. After all, drugs, vaccines and medical devices are not just commodities like an iPod or a designer dress.

10.2 Knowledge for All

The new millennium represents a unique opportunity to ponder the past and fix the future. It began with a sense of excitement and opportunity which, based on the technological advances developed in the last decades of the twentieth century, would make it possible for all in the new century to enjoy a better and more healthy life. The information technology revolution had not only increased the capacity of humans to communicate instantaneously around the world, but also to create and manage knowledge with the same speed. The race to sequence the human genome, and the ultimate triumph of the public sector genome project over the private-for-profit effort to compile and sell the information encoded in our DNA, insured that the complete genome data was deposited in the public domain, where it represents a veritable storehouse of knowledge for all to use. It succeeded because exceptional new technology for rapidly sequencing large chunks of DNA permitted the assembly of a virtual team around the world, managed by a core and visionary group supported by public funds, with the access to cutting edge informatics applications and the capacity for centralized data analysis and publication in real-time. Fortunately, there was both determination and resources to insure that this information would remain a global public good. This should be a paradigm to emulate, but it does not actually happen very often.

10.3
Knowledge Sequestration

New knowledge is the product of effort and innovation, but it is not necessarily available unless disseminated and it is not useful unless it is applied. New knowledge is also the currency of academia and the private sector, and therefore dissemination and application typically occur under a set of constraints. For example, academic career advancement depends on scientific productivity and the publication of new information and insights in journals of the highest quality possible. Information is, therefore, generally not released in real-time, but rather is contained within the laboratory and the institution until sufficient to warrant publication or a public presentation. Information is released in packets over time, often without the details in methodology to allow others to build on the report. When of a particularly innovative or breakthrough nature, may be kept from public disclosure until a patent application has been filed, although the introduction in 1995 of provisional patent applications in the United States has generally eliminated the need for such delays since a provisional patent application can consist of just the manuscript, a cover sheet and a short statement of the scope of protection that will be claimed in the eventual utility application. The private sector pharmaceutical and biotechnology industries also control the release of the information they generate, and because career paths are not so dependent on publication as they are in academia, it is easier to restrict submission of papers. While delayed publication is not necessarily practiced, the time to dissemination may be quite long if it stands to financially benefit the company. This is what we mean by knowledge sequestration.

When there is patent protection for IP, knowledge sequestration may be imposed in other ways. For example, the patent holder could give permission for the IP to be used in research, but as soon as there is the hint of product development the IP holder can control the application of the new findings to the development side of R&D. This hold can only be more complicated when there are a series of stacking patents involved, increasingly the case, since there must be an agreement drawn with all of the relevant stake holders. It is not impossible to get beyond these barriers, however, and there are examples of new public–private partnerships (PPPs) for product development that have successfully managed these hurdles when they relate to priority issues of global health importance and an organization with deep pockets, such as the Gates Foundation, is involved in funding the PPP.

10.4
New Research Models

Research for most of the twentieth century centered around particularly gifted, or occasionally extremely lucky investigators, who organized a laboratory funded by public, institutional or private money based on defined research projects. At the basic end of the spectrum, research was hypothesis driven, or at the least hypothesis testing, and designed to develop fundamental knowledge about how things worked, whether in the natural or the life sciences. In health and biomedical research, we became increasingly reductionist, honing down on cells, and then subcellular structures, then molecules, then genes and so on. The laboratories involved were run by a principal investigator who assumed the responsibility for directing the work that was, by and large, carried out by postdoctoral and predoctoral students, occasional undergraduates, and technicians and various assistants. Large laboratories could be very large indeed, including large amounts of money and increasingly expensive instrumentation. Success was going from grant to grant, with publications on the findings being the essential proof that progress was being made to the peer reviewers who made judgments on who was to be funded and how much they were to receive. Researchers moved up the academic ladder, based on grants received and papers published. It was no secret that the culture was dominated by a 'publish or perish' mentality, and successful investigators got grants, published papers, got more grants and were promoted. To make it all work, funding agencies grew their bureaucracy to administer the process, and created a breed of very committed and very knowledgeable science administrators who ran the enterprise. However, to make advances that actually improved the health of people (or animals, or plants) basic research findings had to be translated into clinical relevance for the biology of cells, organs, individuals and populations, and products that could be delivered and could make a difference had to be commercialized. To connect upstream research to downstream product development required a different set of perspectives and skills. Academic researchers were rarely interested in determining how to produce a drug at scale, and how to package and deliver it in way that was both safe and effective. This R&D part of the process, in essence converting a chemical that could just as easily kill as cure you into a useful drug, has become the purview of industry. Universities might hold the patents for chemicals or drug targets, but it takes industry to make the products that people can use.

How has the research model changed and can it facilitate the hand-over from basic to applied research? For one, industry is doing more and more basic research in-house, and academic scientists are increasingly branching out to form small biotechnology companies. Second, sources of funding are increasingly coming to academic laboratories from industry, with a focus on areas already identified as being of interest to the companies. Third, funders interested in research for product development for neglected diseases, such as the Gates Foundation, are using an industry paradigm for milestone-based research support. Nowhere is this better exemplified than in the Gates Foundation's Grand Challenges in Global Health Program. Finally, because research is becoming increasingly interdisciplinary, multi-investigator, and more oriented towards translation of basic findings for clinical use and the integration of social and biomedical science, there is a better opportunity for academic institutions to recognize accomplishments in applied research and create new pathways for academic advancement. No longer can the academic credo be summed up simply as 'publish or perish'. It is essential that academic institutions and their faculty understand that the role of such institutions – at least in the field of health – is the generation, dissemination, translation, application, implementation and evaluation of knowledge for the benefit of the global population. The vision must be one of education, research and service in a fiscally responsible manner, and not just a financial return on institutional investments in its research capacity.

10.5 Changing Role of Patents in Academic Research

Historically, there was a clear distinction between the roles of academic scientists and corporate scientists in the discovery of new therapies and prophylaxes – academic scientists elucidated the underlying mechanisms of disease, and corporate scientists applied this knowledge to create new interventions and preventatives. Academic scientists therefore rarely entered the world of patenting. When they did, they either did it on their own account, owning and paying for the patents themselves, or the patents were owned by the government agencies which funded their work. These governmental funding agencies had the responsibility for developing the patents, but had cumbersome bureaucracies and frequently adopted well intentioned policies that had the unfortunate side-effect of actively discouraging corporate investment in developing these patents.

In the late 1970s a new paradigm started to emerge.

First, the nature of science changed with the development of recombinant DNA techniques, and the explosion of cellular and molecular biology. The creation of monoclonal antibodies is an example wherein the results of academic research could be more immediately translatable into therapeutic modalities.

Second, a realization dawned that the public good would be better served by policies that encouraged closer collaboration between academic scientists and their counterparts in industry. Institutional Patent Agreements were implemented first by the US Department of Health, Education and Welfare and then by the National Science Foundation, which allowed a significant number of academic institutions to own the patents that resulted from the work these two agencies had funded if the institutions created the infrastructure to commercialize the work in a 'timely' fashion.

The benefits of this 'test marketing' of institutional ownership and management of IP were immediately apparent and were extended to all federal funding agencies and made a general right with the passage of the Bayh–Dole Act of 1980. The amendments to the Stevenson–Wydler Act in 1986 extended many of the same opportunities to US Federal Laboratories.

10.6 Triple Helix and Economic Impact of the New Paradigm

Etzkowitz coined the term the 'Triple Helix' to describe the complex interplay between government, academia and industry that has resulted from the combination of changes in science and changes in public policy in the United States since 1980. There are two clear consequences of these new interactions:

- Diversified high-technology clusters have grown up round all the major research universities and a number of federal laboratories in the United States.
- US$ 40 billion of economic activity and 270 000 jobs has been attributed to the results of academic technology transfer.

10.7
Contributions of Public Sector Research to New Drug Discovery

Nowhere have the benefits of the 'Triple Helix' been more broadly seen than in the discovery of new drugs and vaccines, although the extent of this impact has not been systematically documented to date. The bright line hand-off of responsibilities in the historical paradigm of academia elucidating disease mechanisms and industry discovering cures is no more. Just as industry is itself investing more in to basic research, increasingly academia is discovering the cures and handing them off to industry to develop them. A study currently under way reveals that since 1980 over 140 drugs, biologics and vaccines have been discovered by US public sector researchers (working in national laboratories, universities, hospitals and not-for-profit research institutes), and have been successfully licensed to companies that have gone on to demonstrate safety and efficacy and have received US Food and Drug Administration (FDA) approval. Public sector researchers in other countries – especially the United Kingdom, Canada and Australia – have made similar contributions to public health.

Public sector research is not driven by the financial imperative to discover cures for large populations and fully 25% of the new drugs discovered in public sector research over this period were to treat orphan disorders – defined in the United States as a patient population of less than 200 000. This has important ramifications for the potential for academic research to discover cures for conditions that only afflict developing countries and that will not be large revenue generators.

10.8
Spread of the New Paradigm Worldwide

Some countries have always had rules that allowed for institutional ownership of inventions, as opposed to ownership by the inventing professor or the funder of the research, generally the government. Those that followed one of these two alternative models have started to adopt the US model. The United Kingdom was the first to follow suit, when the Thatcher government eliminated the National Research and Development Corporation as the patenting and licensing agent for British universities. Over the second half of the 1990s and accelerating into the first decade of the twenty first century other countries have adopted the US model. Outside of the United States, academic institutions do not generally have the substantial discretionary funds that US institutions have access to from tuition, philanthropy and indirect costs, so funding the investments in staff and patenting needed to capitalize on their inventions often requires waiting for governmental initiatives.

10.9 Role of Patents in the Academic Mission

10.9.1 Patents and Publishing

Historically, patenting and publishing were thought to be in conflict – a scientist can only do one thing at a time, and if forced to prioritize most academic scientists would still choose to publish first and patent second. US patent law allows an inventor 1 year after they publish their work to apply for a patent, so US rights would still be available to scientists making this choice, but outside of the United States, in general, an 'absolute novelty' rule applies and no rights would be available outside the United States. Where these represent 'markets' for a product there may still be pressure to withhold publication until filing for patent protection has been accomplished.

The landscape changed with the General Agreement on Tariffs and Trade treaty, when the United States introduced provisional patent applications. These can range in sophistication from a cover sheet added to a manuscript or grant application all the way up to a traditional utility patent application. They give the applicant 1 year to file a full utility application. Provisional patent applications eliminated the conflict between publishing and patenting, although this is probably still somewhat mysterious to the majority of academic researchers.

10.9.2 Patents and the Power to Dictate the Terms of Development

In some quarters, patents are regarded as antithetical to the open collaborative nature of science. This view is mistaken. Patents are themselves a form of publication – all patents now are published 18 months after their initial filing – and the very reason that patents were enshrined in the US Constitution was to encourage inventors to disclose their inventions so that others could build on them and make further advances.

It is true that after a patent has issued on an invention even performing scientific research on the invention is an infringement of the patent. While many scientists think there is a 'research exemption' the only research exemption under patent law is to the right to make and test a drug in order to be able to submit an Abbreviated New Drug Application to the FDA so as to be in a position to start sales of the drug the first day after the patent has expired.

However, there is a convention amongst academic institutions not to enforce patents against each other and to seek licenses from each other for scientific research. Academic institutions granting exclusive licenses to companies for commercial development always reserve the right for themselves, and hopefully for other academic institutions, to perform further research without taking a license and paying fees.

Rather, patents pertain solely to the commercial development of a scientific discovery. They allow the holder of the patent to control the way in which it is developed – or indeed to ensure that it is not developed at all.

10.9.3
Patents versus Licensing

That control is exercised through licensing. A patent holder can decide to issue an exclusive license to develop a new product to just one company. In this case the price for the product will probably be high. Alternatively, the patent holder can decide to allow multiple licenses to the drug, in which case prices will likely be lower. The risk in this approach is that no company will be prepared to make the substantial investment needed to prove that the drug is safe and effective for the first time.

For diseases that strike both the developed and the developing world, it is, of course, possible to do both, carving up the world into two (or more regions) and licensing exclusively in the developed world, where monopoly profits can repay the cost of development, and licensing non-exclusively in the developing world so that the product can be sold at generic prices or cost plus some level of profit.

Over the past 5–6 years, the pharmaceutical industry has been forced by the sheer weight of public opinion to accept just such a 'two-tier' pricing scheme for selected drugs, particularly anti-retrovirals, where it controlled worldwide patents. It is incumbent on the public sector when it invents, patents and licenses a drug or vaccine to ensure that its licensing practices are designed to generate two-tier pricing from the outset. Industry is now starting to embrace this approach.

For diseases that afflict only the developing world, it may be necessary to seek philanthropic funding for clinical development (perhaps through one of the public–private drug development partnerships) and to keep exclusive rights available to induce a developing country drug company to invest in developing a manufacturing process, retaining the right to seek a second source to 'keep them honest' and ensure that prices stay low.

10.10
Managing IP in Research Networks

Global research and development networks are critical in developing drugs to treat the diseases of the developing world. Only a global network can bring together all the resources needed to attack a disease effectively:

- Sophisticated academic laboratories in the west, and increasingly in some middle-income nations, to secure government and philanthropic funding and to lead the scientific effort.
- Sophisticated academic laboratories in the heavily affected countries to translate the first groups' discoveries into therapeutic modalities that will be accepted by society and are practicable under local conditions.
- Clinical centers in the affected countries to provide access to patients for clinical development.
- Manufacturers in developing countries to manufacture and distribute products of high quality, up to international standards, and at affordable prices for the developing world.

Local manufacture can be particularly important, because many pharmaceutical companies only secure patents in the United States, Europe and Japan, and while there may be freedom to sell in the developing world, there will likely not be freedom to manufacture in these countries to supply the developing world. Having a manufacturer outside the high income regions can obviate the need to obtain licenses in order to manufacture for export and the costs of manufacture, including production facilities, can be substantially reduced as well.

It is likely that for some years to come such a research network will be based primarily on the sophistication and capability in academic institutions in the north, with increasing contributions over time from institutions in middle- and lower-income countries. It would be valuable at the outset of such a network to designate one of these institutions as the IP manager for the consortium. Each member of the research consortium should, in turn, designate an IP manager who will represent them on an IP committee for the group which will take decisions on IP matters

The lead institution should be charged with developing a consortium agreement, which all members of the consortium will sign, that spells out the expectations of all the parties with respect to:

- How the IP will be managed
- How it will be paid for
- How it will be licensed
- Whether it is intended that IP generate income for the consortium
- If so, how any income will be distributed

Another topic that should be covered in the consortium agreement is whether any of the parties own background rights – patents that will need to be licensed to conduct the program, and to make and sell any resultant products. That party should be prepared to grant a research license to the consortium members to carry out the project and should specify under what terms they will grant licenses for commercial production.

It is important that there be regular meetings of the IP committee by conference call and, if possible, at least once a year in person. This could be organized at the same time as an annual scientific program review meeting, or it could be held at the Annual Meeting of the Association of University Technology Managers or another suitable gathering of IP leaders.

10.11 Conclusions

Global public goods are an important consideration for academic institutions involved in structuring patenting and licensing agreements. This pertains most specifically to research funded by the public sector, but should also become a part of all patenting and licensing decisions, regardless of the source of funds. Changes in the nature of research funding, including multisectoral support from the public and private sectors, and in the nature of research teams and their increasingly interdisciplinary constitution, including both other academic institutions and industry, require new vision in the management of IP and in setting goals for institutional and professional attainment and career pathways. New operating procedures will need to be crafted for the new networking and multi-institutional format of research and the management of IP on behalf of the public.

Part III
New Frontiers

Technology Transfer in Biotechnology. A Global Perspective.
Edited by Prabuddha Ganguli, Rita Khanna, and Ben Prickril

ISBN: 978-3-527-31645-8

11
Biotechnology in the Midst of a Global Transformation

G. Steven Burrill

11.1
Introduction

Biotechnology has reached a tipping point. Not only has it transformed into a massive global enterprise, but the way in which businesses are creating value is transforming too. This new environment presents both opportunities and challenges for the biotech industry.

The global economy is in the midst of radical transformation with far-reaching and fundamental changes in technology, production and trading patterns. Rapid technological change continues to impact on how individuals, businesses and communities interact with each other and with governments. Global security is being reshaped as the international community responds to the ongoing threat of international terrorism, conflict and the challenges of ending world poverty. The pressures that economic and population growth are placing on the Earth's natural resources and climate is increasingly apparent, presenting an urgent need for international cooperation.

The biotech company that we start today operates globally. It is not constrained by physical location; science is available globally, investment capital and resources can be accessed globally as can data on a 24/7 basis by connecting with the web. We have instantaneous communications globally. Patents and intellectual property protection are global issues. This transformation will present both challenges and opportunities for the biotech industry and individuals, businesses and communities all need to be ready to respond to the changing global environment.

Within this milieu, biotech is undergoing its own global transformation before our very eyes. The industry is no longer centered in the United States and Europe: its maturity now means that competition for resources (human and technological) is increasing as the number of countries supporting viable life sciences industries grows. Nearly every part of the world is looking to its future dramatically affected by biotechnology.

Biotechnology research, development and commercialization are no longer limited to the domain of industrialized nations – biotechnology is now truly global.

Technology Transfer in Biotechnology. A Global Perspective.
Edited by Prabuddha Ganguli, Rita Khanna, and Ben Prickril

ISBN: 978-3-527-31645-8

This dramatic changing landscape is evidenced by the fact that a decade ago 75% of the biomedical citations in professional scientific journals were linked to US scientists, today that figure has dropped to less than 30%. In a world where advanced knowledge is widespread and low-cost labor is readily available, US advantages in the marketplace and in science and technology have begun to erode. A recent report[1] cites the fact that for the cost of one chemist or one engineer in the United States, a company can hire about 5 chemists in China or 11 engineers in India. 'Chemical companies closed 70 facilities in the United States in 2004 and have tagged 40 more for shutdown', the report says. 'Of 120 chemical plants being built around the world with price tags of US$ 1 billion or more, one is in the United States and 50 in China.'

China and India, for example, are becoming attractive alternative locations for drug development. They provide the Western world biopharmaceutical industry with the opportunity to develop drugs at lower costs. As barriers to long-distance national and global transactions have fallen through advances in technology and logistics such as the growth of the Internet and overnight package services, it has become increasingly possible to relocate operations such as research, product development and manufacturing to countries and regions with relevant expertise and lower costs. Meanwhile, more companies are discovering that they too can strategically invest their capital across borders. This process continues to accelerate, fueled by an ever-increasing sense of urgency to be on the cutting edge.

The global transformation is not merely referring to geography. It is occurring pan-industry and across technology, and innovation spurring. We have begun to understand disease 'globally' (i.e. from a systems point of view – from gene or single nucleotide polymorphism to protein to networks to disease). In parallel, and conversely, the notion of 'one size fits all' is being replaced by medicine targeting the individual. We have great expectations for the era of personalized medicine, which promises to catalyze a major transition in health care. Factor in a greater understanding of disease on a 'systems' level and the promise for the future of personalized, predictive and preventative health care is compelling – earlier and more precise diagnoses, treatments tailored to the individual, reduction of side-effects and adverse reactions to drugs, breakthroughs in treatment, and ultimately prevention of major diseases such as cancer, diabetes and Alzheimer's. To achieve these lofty goals requires a global effort. We have to take the international cooperation that galvanized in the genomics arena throughout the Human Genome Project to the next level. A global effort is certainly needed to adequately prepare for diseases in which the majority of world's population has no innate immunity – no one country is up to this task where disease 'knows no borders'.

1) Committee on Science, Engineering and Public Policy (2007) *Rising Above the Gathering Storm: Committee on Prospering in the Global Economy of the 21st Century: An Agenda for American Science and Technology*, National Academies Press, Washington, DC.

11.2 Healthcare

Human healthcare costs in the United States continue to rise, pushing it to the US$ 2 trillion level. Although drugs constitute less than 10% of the annual healthcare bill, they are perceived by many as a major part of the problem. The Democrats gained control of the House and Senate riding a platform to get costs under control by emphasizing generics and greater oversight on drug spending by Medicare.

Part of the solution is seen in the convergence of information technology and genomics, smarter drug delivery and 'labs' on a chip, which are leading us down a path towards targeted personalized medicines – away from solely concentrating on 'blockbusterology' – and spotting 'early warning' signs of impending health problems.

The industry continues to wrestle with improving its productivity and ensuring product safety. These issues kept drug approvals down and in an effort to improve the situation, the US Food and Drug Administration (FDA), which finally got a full-time Commissioner, turned up the volume on its Critical Path Initiative agenda, pushing the industry towards pharmacogenomics and theranostics.

In response, pharmaceutical companies are casting off more of their fixed costs to focus instead on core competencies in late-stage development, distribution and marketing. This is pushing them increasingly toward a more disaggregated model, relying on biotech firms that will continue to do research and early development and contract research organizations to do the testing and others offer to sell or co-market their drugs.

Drug developers are moving from the 'one-size-fits-all' to individually tailored medicines. Drug companies are already using genetic/proteomic and other screening tools to figure out what patients are best suited for which drugs. The theranostics era is upon us. The movement is most obvious in cancer, where drugs have been approved specifically for a subpopulation of patients displaying particular genotypes. While the business model for diagnostics has been maligned as a commodity, those tests that direct the course of expensive therapy can command high prices. Genomic Health's Oncotype Dx, for example, costs around US$ 3000. It lets a woman with a certain kind of breast cancer know whether her cancer has a 3 or 33% chance of recurrence – information that many physicians and their patients want to know when deciding whether or not to undergo the suffering of chemotherapy.

Regulators are preparing for a raft of new drug applications likely to come with companion diagnostics, based on everything from simple genotype measurements to more complex assessments of gene activity, protein expression, and other factors. As just one example, Pfizer has teamed up with Monogram in a large clinical trial of an anti-HIV drug that prevents the virus from gaining access to patients' white blood cells. Versions of HIV differ in which protein they use to infect cells – Monogram's test identifies this protein and so helps Pfizer predict how well patients will respond to the drug.

Medical advocacy and medical philanthropy are also changing the environment. Philanthropists are not just satisfied in donating money. They view themselves as ultimately responsible for the performance of the organizations in which they invest, holding them accountable for achieving benchmarks of progress in both process and science along the way to the ultimate outcome of improving lives of patients and society. The spin-off of innovation – potential therapies and vaccines – from medical foundations could be the equivalent to the influence that NASA's space research program had on innovation that led to many of the indispensable products that we use every day.

11.3 Financing Environment

While biotech's performance in the capital markets was choppy throughout the year at the mercy of prevailing macro-economic forces, concern for Iraq, elections/politics, and healthcare cost increases, it was a big year for biotech/life sciences fund raising. Financings and partnering deals brought in a record US$ 47 billion for US companies with over US$ 27 billion through financings and US$ 20 billion in partnering. The industry's collective market cap finished the year back where it started, at US$ 490 billion. A contributing factor was the macro-markets, which took their toll on investor confidence and saw them gravitate away from biotech into other industrial sectors including energy. As a result, the NASDAQ, Dow Jones Industrial Average and the pharma indices outperformed the Burrill Biotech Select Index by a wide margin. Big pharma stocks did very well as Merck and other big pharmas recovered from their woes. In contrast, biotech's big guns – Amgen and Genentech – slid after a couple of very good years. Interestingly, we saw some stellar gains by individual large cap biotech companies such as Gilead and Vertex Pharmaceuticals that were balanced out by companies that crashed and burned.

It was the mid-cap and small-cap companies that carried the load in 2006. The Burrill Mid-Cap Biotech Index was up 10% and the Burrill Small-Cap Biotech Index performing better, up 14%, and these increases mirrored the Dow, which was up a respectable 16%, and NASDAQ up 10% for the year.

The ongoing transition to a more personalized medicine world was not lost by Wall Street. Companies specializing in molecular diagnostics, biomarkers, genotyping assays, and other fields, all received positive investor attention and returned double-digit gains for the year.

Several billion dollar debt financings were completed in 2006, particularly in quarter (Q) 4 2006, so it was no surprise that the US$ 862 million collectively raised by 20 companies in Q1 2007 was 50% down from the Q4 2006 total. Leading the way was specialty pharmaceutical company, Alpharma, which raised US$ 300 million and ViroPharma will use its US$ 250 million debt financing for strategic investments and/or acquisitions of products (Table 11.1).

Table 11.1 Selected convertible debt offerings during Q1 2007.

Company	Amount (US$ million)
Alpharma	300
ViroPharma	250
Isis Pharmaceuticals	125
Encysive Pharmaceuticals	60
Vion Pharmaceuticals	60

Table 11.2 Selected secondaries during Q1 2007.

Company	Amount (US$ million)
Inverness Medical Innovations	273
Medarex	152
Vanda Pharmaceuticals	120
Myriad Genetics	105
Allos Therapeutics	54
Kosan Biosciences	45
BioMimetic Therapeutics	43
Nastech Pharmaceutical	42
Cyclacel Pharmaceuticals	36
Repros Therapeutics	36

Table 11.3 Selected PIPES during Q1 2007.

Company	Amount (US$ million)
Neose Technologies	43
Iomai	32
Ziopharm Oncology	31
Cleveland Biolabs	30
Icagen	22

Although the same number of follow-on deals was completed in Q1 2007 as the previous quarter, the total amount raised of US$ 1.09 billion was 60% down from the US$ 2.7 billion that biotechs raised in Q4 2006 (Table 11.2)

Financings through private investment in public equities (PIPEs) picked up the pace once again in Q1 2007 generating 60% more than in Q4 2006 (Table 11.3).

11.4 Signs of Growth in 2007

The collective market cap of the 360 public biotech listed on the NASDAQ and Amex, and monitored by Burrill & Company, soared above US$ 500 billion for the first time in its history. The new mark was set on 17 April 2007 and since that date it went on to close the month at US$ 507 billion. This growth is even more remarkable when you take into account the fact that we have seen several leading biotechnology companies, with multibillion dollar market caps, being acquired by big pharma and their market cap value removed from the industry's collective total. A good example of this is MedImmune, which contributed around US$ 13 billion to the market cap total. With AstraZeneca announcing its intention to acquire the company for US$ 15.2 billion, the industry will need to continue on its upward path to compensate.

11.5 Mergers and Acquisitions Now Part of the Industry Scene

Pharma/biotech and biotech/biotech consolidation continues to be red hot as pharmas looked increasingly outside for their organizations pipeline and access to innovation. The mergers and acquisitions trends, that have been hot in 2005 and 2006 in biotech land, are not slowing down with pharma desperate to access pipeline and innovation. Both big pharma and big biotech are competing for companies with advanced product pipelines, as well as important land grabs of technology such as the US$ 1.1 billion acquisition of Sirna by Merck announced in November 2006. It is likely that we will consistently see these types of billion dollar transactions as pharma and biotech jockey for best-of-breed technology and market leadership in the almost US$ 650 billion worldwide pharmaceuticals market.

11.6 Deal Making Slows Slightly

Although partnering cooled off somewhat at the start of 2007, the biotech industry's record for deal making remains outstanding and indicates healthy growth as companies continue move their products forward, garnering the necessary resources and cash from strategic partners. The US$ 4.4 billion in announced transaction values was 40% down from the US$ 7.3 billion generated in Q4 2006.

Interestingly, it was not big pharma that headed the deal makers. Genentech entered into an exclusive worldwide license agreement with Seattle Genetics. The deal involves the development and commercialization of SGN-40 – a humanized monoclonal antibody currently in phase I and phase II clinical trials for multiple myeloma, chronic lymphocytic leukemia and non-Hodgkin's lymphoma.

Table 11.4 Selected partnering transactions during Q1 2007.

Biotech	Pharma/biotech	Description	Estimated value (US$ million)
Seattle Genetics	Genentech	SGN-40 (a humanized monoclonal antibody)	860
Cytokinetics	Amgen	cardiovascular therapeutics	675
Anacor Pharmaceuticals	Schering Plough	topical anti-fungal therapy	625
OSI Pharmaceuticals	Lilly	diabetes program	385
Palatin Tech	AstraZeneca	melanocortin receptors	310
Avalon Pharma	Merck	inhibitors for a cancer target	200
BioInvent	Genentech	antibodies for cardiovascular diseases	175
KV Pharmaceutical	Vivus	estrogen transdermal spray	150

Seattle Genetics will receive an upfront payment of US$ 60 million, potential milestone payments exceeding US$ 800 million and escalating double-digit royalties on annual net sales of SGN-40. Genentech also struck another antibody deal with Swedish company BioInvent to develop and commercialize its proprietary antibody candidate, BI-204, for the potential treatment of multiple cardiovascular conditions. Genentech will make an upfront payment of US$ 15 million and potential milestone payments of up to US$ 175 million as well as royalties on sales in North America.

Amgen was also in deal making mood entering a strategic collaboration with Cytokinetics on small-molecule therapeutics that activate cardiac muscle contractility for potential applications in the treatment of heart failure. Cytokinetics receives a non-refundable up-front license and technology access fee of US$ 42 million. Amgen also purchased 3 484 806 shares of Cytokinetics common stock at US$ 9.47 per share and an aggregate purchase price of approximately US$ 33 million. In addition, Cytokinetics may be eligible to receive milestone payments of up to US$ 600 million (see Table 11.4).

11.7 Weak Initial Public Offering (IPO) Market Does Not Deter Hopefuls

The IPO market languished once again; only 19 biotech IPOs managing to get done in the United States in 2006, two more than in 2005. To ensure that the deals were completed, almost all the offerings priced at the low end of their pricing ranges or considerably lower.

Table 11.5 Biotech IPOs completed in Q1 2007.

Company	Ticker	Issue price	Amount raised (US$ million)	Price (3 October 2006)	Change since IPO (%)	Market cap (US$ million) (3 October 2006)
Rosetta Genomics	ROSG	7	30	6.79	−3.00	76
Optimer Pharmaceuticals	OPTR	7	56	9.71	38.71	211
3S Bio	SSRX	16	123	11.02	−31.13	157
Synta Pharmaceuticals	SNTA	10	50	8.07	−19.30	273
Molecular Insight Pharmaceuticals	MIPI	14	70	11.88	−15.14	292
Oculus Innovative Sciences	OCLS	8	24	5.95	−25.63	70
Average		10.33	58.83	8.90	−9.25	180

Only one of the six biotech companies completing IPOs in Q1 2007 was above water by the end of the quarter (Table 11.5). Optimer Pharmaceuticals bucked the trend of this group on the strength of news that it was advancing its lead antibiotic drug candidate Difimicin for the treatment of *Clostridium difficile*-associated diarrhea into phase III clinical studies. However, this did not deter 14 companies to join the IPO runway.

11.8
Venture Capital: Deals Continue to Flow

The amount of venture capital generated by biotechs in Q1 2007 was up 33% over the Q4 2006 period and 100% greater than the comparative period in the previous year (Tables 11.6 and 11.7). The 56 reported deals in the quarter averaged about US$ 21 million per investment – an amount that has held steady for several quarters.

EUSA Pharma, a new transatlantic specialty pharmaceutical company that focused on developing and marketing products for the hospital market both in Europe and the United States, acquired OPi (Lyon, France) using the US$ 175 million that was invested into the company. OPi has four products in the market, a further two products in clinical development and a fully human anti-interleukin-6 monoclonal antibody in preclinical development. Targanta Therapeutics, which is developing anti-bacterial drugs, raised US$ 70 million in a Series C venture round. The company intends to utilize the capital to prepare and submit an New Drug Application for its lead product, oritavancin, for the treatment of complicated skin and skin structure infections.

Table 11.6 Selected venture financings during Q1 2007.

Company	Amount (US$ million)
EUSA Pharma	175
Targanta Therapeutics	70
Ception Therapeutics	63
Omeros	63
Phenomix	55
Microbia	50
Intarcia Therapeutics	50
MAP Pharmaceuticals	50
Cylene Pharmaceuticals	44
TransOral Pharmaceuticals	40
Fluidigm	37
Epiphany Biosciences	36
Sequoia Pharmaceuticals	35
Proacta	35
BiPar Sciences	35

Table 11.7 US biotech financings (US$ million).

	2005	2006					2007	Total
	Total	Q1	Q2	Q3	Q4	Total	Q1	
Public								
IPO	819	252	269	49	350	920	353	353
secondaries	4194	1522	1091	419	2734	5766	1055	1090
PIPEs	2376	1042	295	480	210	2027	333	333
debt	5565	5421	6447	373	1737	13978	862	862
Private								
venture capital	3518	734	1352	1100	1050	4236	1401	1401
other	1114	115	98	90	122	425	267	267
Total	17586	9086	9552	2511	6203	27352	4241	4316
Partnering	17268	6436	1809	4218	7333	19796	4383	4383
Total	34854	15522	11361	6729	13536	47148	8624	8699

12
Technology Transfer in Biotechnology: International Framework and Impact

Carlos M. Correa

12.1
Introduction

'Biotechnology' encompasses a broad set of techniques, some of them simple, others complex, such as the genetic manipulation of living organisms. Farmers have exchanged over the centuries technologies and breeding methods to improve agricultural production. Such exchange of knowledge and materials has been conducted on an informal basis,[1] and has enormously benefited both developed and developing countries.[2]

A different picture arises when modern biotechnology is at stake, particularly the applications of modern biotechnology, where a sharp north–south divide can be found. Different factors limit the transfer of modern biotechnology to developing countries. On the supply side, some technologies, such as those relating to the genetic transformation of plants, are controlled by a handful of firms ([2], p. 148). The globalization of the economy and the extensive liberalization of markets allow technology owners to directly export innovative products without resorting to foreign direct investments or licensing. Some innovative firms are particularly reluctant to transfer technology that may help a potential licensee to become a competitor in the global market or when they perceive a risk of 'leakage' of knowledge leading to imitation [3]. Moreover, in the case of agrobiotechnologies, developments have 'been limited to a small number of traits of interest to commercial farmers .... Very few applications with direct benefits to poor consumers or to resource-poor farmers in developing countries have been introduced' ([4], p. 931).

1) Transfer of technology can take place through a diversity of means, both formal, (e.g. licensing agreements) and informal (e.g. subcontracting, acquisition of machinery). See, e.g. [1], p. 16.
2) In recognition of such contribution, the Food and Agriculture Organization of the United Nations (FAO) Resolution 4/89 introduced the concept of 'Farmers' Rights' further defined by FAO Resolution 5/89 as '[R]ights arising from the past, present and future contribution of farmers in conserving, improving and making available Plant Genetic Resources, particularly those in the centers of origin/diversity ...'.

Technology Transfer in Biotechnology. A Global Perspective.
Edited by Prabuddha Ganguli, Rita Khanna, and Ben Prickril

ISBN: 978-3-527-31645-8

There are also factors on the demand side that limit technology transfer. Technology is not just information that can be easily communicated: its transfer requires capacity to learn and to incorporate it into the recipient's production system. Technology transfer requires both a party willing to share its knowledge and a recipient able to absorb and apply it. This is far from being an automatic and costless process, particularly due to the 'tacit' nature of much of the knowledge involved (see, e.g. [5], p. 17).

Given the very asymmetric distribution of R&D capacity in the world (see, e.g. [6]), developing countries are strongly dependent on foreign technologies to advance in their development strategies. Despite the advances made in some developing countries (see, e.g. [7]), this applies to most fields of technology, including biotechnology.

In this chapter, I (i) examine the international legal framework for technology transfer, particularly developments within the World Trade Organization (WTO), (ii) consider the case of Least Developed Countries (LDCs), (iii) discuss some policies to foster technology transfer, and (iv), finally, refer particularly to the case of biotechnologies for agriculture and other industries.

12.2
International Framework for Technology Transfer

Given developing countries' needs to get access to foreign technologies, their interest in the development of an international framework that facilitates such transfer is not surprising. In the 1970s, where mature technologies (to be applied in protected markets) were abundantly transferred, developing countries unsuccessfully proposed the adoption of an International Code of Conduct on Transfer of Technology.[3] The draft Code reflected concerns about the *conditions* (notably restrictive practices and guarantees) under which technology transfer took place, rather than about the availability of and access to technologies developed in the industrialized world. However, in the 1990s, especially after the tightening of intellectual property (IP) rights ensuing from the WTO's Trade Related Aspects of Intellectual property Rights (TRIPS) Agreement,[4] concerns shifted to the issue of access to technologies as such. In a context of growing liberalization of their economies, developing countries found more it difficult to obtain the up-to-date technologies they needed to compete in a globalized market.

Against this backdrop, a group of developing countries submitted, in September 2001, a proposal to the WTO, for the establishment of a Working Group for the study of the inter-relationship between trade and transfer of technology. Such a Working Group was established by the Fourth WTO Ministerial Conference

3) See UNCTAD TD/COT TOT/47 (1985). The Code of Conduct was developed under the auspices of the United Nations Conference on Trade and Development (UNCTAD).

4) Available at www.wto.org.

(Doha, Qatar) with a mandate to examine the relationship between trade and transfer of technology, and to make recommendations on steps that might be taken within the mandate of the WTO to increase the flow of technology.[5)]

The Working Group examined, *inter alia*, the provisions relating to transfer of technology in the various WTO Agreements. A paper submitted by a group of developing countries in October 2002 highlighted that, in most cases, such provisions – mainly concerning technical assistance[6)] – contained only 'best endeavor' commitments, but not mandatory rules,[7)] and that no significant benefits could be obtained therefrom. After 5 years of its establishment, the Working Group has been unable to devise a general strategy or suggest concrete mechanisms to respond to developing countries' concerns.[8)]

The Working Group did not directly deal with issues under the TRIPS Agreement. This has been a serious limitation to its work, since that Agreement sets out the rules for the appropriation of IP assets. In accordance with Article 7 of the Agreement:

> The protection and enforcement of intellectual property rights should contribute to the promotion of technological innovation and to *the transfer and dissemination of technology*, to the mutual advantage of producers and users of technological knowledge and in a manner conductive to social and economic welfare, and to the balance of rights and obligations [emphasis added].

Although Article 7 indicates that Members should implement their obligations under the Agreement in a way that contributes to 'the transfer and dissemination of technology', the TRIPS Agreement essentially is a charter for the protection of IP rights and only contains a few elements that may effectively contribute to that end (see, e.g. [8]). While it may be argued that the strengthening of IP protection in developing countries may encourage the transfer of technology thereto (see, e.g. [9], pp. 41–74), the increased power it confers on titleholders essentially leaves at their discretion whether to transfer or not the technologies they possess, and to determine the price and other conditions thereof. Although there is no conclusive evidence of an increase in the flows of production technologies to developing countries, there has been an impressive rise in global royalty payments.

5) See WTO document WT/GC/W/443, available at www.wto.org.
6) See, for instance, Article 9 of the Agreement on Sanitary and Phytosanitary Measures and Article 11 of the Agreement on Technical Barriers to Trade.
7) See WTO document WT/WGTT/3/Rev.1, available at www.wto.org.
8) See, e.g. WTO document WT/WGTTT/W/67, May 2003, Communication from Cuba, India, Indonesia, Kenya, Pakistan, Tanzania and Zimbabwe. It suggested that the Working Group examine the need for and desirability of internationally agreed disciplines on transfer of technology with a view to promote trade and development, and come up with appropriate recommendations regarding possible internationally agreed commitments in the field of transfer of technology to developing countries and LDCs. This suggestion, however, has not been pursued by the Working Group or in other fora.

They grew from US$ 61 billion in 1998 to US$ 120 billion in 2004, the United States being the main beneficiary thereof.[9)]

The control of restrictive practices in licensing agreements – permissible within certain limits (see, e.g. [10]) under the TRIPS Agreement – may facilitate access to technology under reasonable commercial conditions. Compulsory licenses may, in turn, allow access to and the use of patented technology on the grounds determined by the national law in accordance with Article 31 of the TRIPS Agreement.

The use of compulsory licenses, however, has been extremely low, except in the United States, where thousands of patents have been subject to such licenses to remedy anti-competitive practices or for government use (see, e.g. [11], p. 8). Only recently have some developing countries started to make effective use of this safeguard (see, e.g. [12]). The deliberate dilution of the possibility of granting compulsory licenses in cases of lack of working of a patent – a possibility that the Paris Convention clearly recognized but which has been contested by the USA under the TRIPS Agreement – has reduced the potential use of such licenses as a means to put into operation patented technologies without the consent of the patent owner. In addition, it is well known that patent documents normally do not contain all the information (especially the know-how) required for a viable production of the protected invention. In some cases the information contained in patent specifications is sufficient to execute the invention by a person ordinarily skilled in the art in a developed country, but insufficient to less knowledgeable people in developing countries.[10)]

12.3 Transfer of Technology to LDCs

According to Article 66.2 of the TRIPS Agreement,[11)] developed Member countries are obliged to provide incentives under their legislation to enterprises and institutions in their territories for the purpose of promoting and encouraging the transfer of technology to LDCs 'in order to enable them to create a sound and viable technological base'. It has been unclear how this obligation was to be implemented, given the generality of the text and the lack of criteria to measure the efficacy of the measures to be adopted. Interestingly, this provision of the TRIPS Agreement is premised on the concept that the implementation of IP protection, as required by the Agreement, is not conducive to the technological development of LDCs and that more flexibility is needed. This is precisely the argu-

9) Based on the World Bank's World Development Indicators (2000 and 2006).

10) For this reason, it has been suggested that the disclosure requirement be more stringent in developing countries so as to increase the informational value of patents in that context ([13], pp. 457–8).

11) Article 66.2: 'Developed country Members shall provide incentives to enterprises and institutions in their territories for the purpose of promoting and encouraging technology transfer to least-developed country Members in order to enable them to create a sound and viable technological base'.

ment that developing countries articulated, during the TRIPS negotiations and that have thereafter reiterated in different fora, such in the context of the 'Development Agenda' submitted by a group of developing countries to the World Intellectual Property Organization.[12)]

At its meeting of September 1998, the Council for TRIPS agreed to put on the agenda the question of the review of the implementation of Article 66.2 and to circulate a questionnaire on the matter in an informal document of the Council. On 19 February 2003, the Council adopted a Decision on the Implementation of Article 66.2 of the TRIPS Agreement, which establishes mechanisms for 'ensuring the monitoring and full implementation of the obligations in Article 66.2', including the obligation to 'submit annually reports on actions taken or planned in pursuance of their commitments' under said article and their review by the Council at its end of year meeting each year.

The issue of transfer of technology to LDCs was also addressed in Paragraph 7 of the Doha Declaration on the TRIPS Agreement and Public health, which reaffirmed:

> ... the commitment of developed-country Members to provide incentives to their enterprises and institutions to promote and encourage technology transfer to least-developed country Members pursuant to Article 66.2.

Though the wording in Paragraph 7 is broad, its inclusion in the Doha Declaration indicates that effective incentives should be granted in developed countries in order to specifically foster the transfer to LDCs of health-related technologies, including pharmaceutical technologies. An interesting aspect of the Declaration is that it refers to 'commitments of developed-country Members', thereby confirming that Article 66.2 does not contain a mere 'best efforts' obligation.

The Decision of 19 February 2003 and the Doha Declaration are steps forward for the implementation of Article 66.2, but concrete measures to facilitate access to technologies by LDCs are still inexistent or insufficient. Given that Article 66.2 belongs to a treaty specifically dealing with technologies protected under IP rights, a logical interpretation is that developed countries are obliged to transfer IP rights-protected technologies and not only those that are already in the public domain.

To what extent could the mechanisms contemplated under Article 66.2 be implemented to foster the transfer of biotechnology? It is conceivable that the incentives that must be provided in accordance with Article 66.2 could be focused on one or more particular areas. For instance, LDCs may be supported to exploit their genetic resources and traditional knowledge through the use of modern biotechnologies. The aim should be to provide them the *tools* to develop their own products and processes, rather than to transfer them finished technological packages.

12) See WIPO document WO/GA/31/11, Annex 5.

12.4 Favoring the Transfer Technology to Developing Countries

There are many factors that crucially influence the creation of an innovation capacity in biotechnology. These include the availability of qualified personnel, good industry–universities linkages, funding, particularly subsidies and venture capital, the legal framework for R&D and the commercialization of innovation (including biosafety regulations and IP rights), national policies to support start-ups, and general industrial and services infrastructure (see, e.g. [14]). Some developing countries have succeeded in the 'development of leading-edge innovations, such as the case of the meningitis B vaccine in Cuba, but most countries have relied on licensing preexisting technology, as was the case for recombinant insulin development in Egypt' [14].

R&D in biotechnology is capital intensive, requires abundant skilled labor and has a long maturity curve. In most developing countries, however, there is inadequate or expensive financing, venture capital is inexistent, and the skilled labor is insufficient and inappropriate. In this context, even if universities or firms can detect useful ideas, they would not have resources and the appropriate infrastructure to implement them. In such a context, moreover, the patent system will not work as an incentive to innovation, and greater reliance on foreign expertise and technologies is unavoidable.

Technology transfer occurs mainly in the context of commercial transactions, but the process is subject to market failures. Often, those who own valuable technologies are unwilling to transfer them as potential licensees might pose, if they learn how to apply them, a competitive threat. In cases where technologies are patented in the jurisdiction where its utilization is sought by local companies or institutions, patents may be an obstacle to technology transfer, unless compulsory licenses are effective in allowing their use (provided, in addition, that the information disclosed in the patent is sufficient for the execution of the invention).

In the case of biotechnology, the relevant knowledge is often 'tacit' and cannot be obtained through access to materials derived from or transformed using biotechnological processes.[13] An active intervention of the technology holder through technical assistance, training and appropriate documentation is, hence, necessary.

A set of policies are needed to remove market impediments and to lower the costs and risks of technology acquisition. Developing such policies is not a simple task, since countries at different development stages benefit from different sources and different forms of international technology spillovers, and economic policies play a significant role in determining both the amount and form of such spillovers [16].

13) A 'technical protection' exists in the case of many products, which makes it difficult or impossible to reverse engineer biotechnological them. See an analysis of the subject in Jullien [15].

A main problem in this area is that there are no internationally agreed rules for facilitating transfer of technology, flexible enough to be adapted to different local situations and sectors, such as biotechnology. As noted by Mytelka [17]:

> While barriers to investment are coming down rapidly and consequently capital is becoming highly mobile, the mobility of other factors of production like labor and technology is becoming increasingly restricted.... This untenable situation is sometimes sought to be defended on the argument that technology is privately held and that therefore, governments cannot make rules relating to transfer of technology. In this context, one cannot forget the fact that IPRs [intellectual property rights] are essentially private rights and that there is a full-fledged WTO Agreement relating to IPRs. One cannot also ignore the fact that there is a demand for international rules on investment in the WTO, though investment is essentially a private sector activity. It cannot be anybody's case that only those topics/subjects/issues where developing countries have to undertake commitments without receiving commensurate benefits, should be brought into the WTO.

One possible approach to this matter is to review the provisions in different WTO Agreements, in order to determine those that hinder transfer of technology to developing countries and to suggest amendments thereto in order to foster such transfer (e.g. changes to the Agreement on Subsidies and Countervailing Measures so as to permit subsidies specifically aimed at promoting the transfer of technology to developing countries and LDCs). A complementary route would be to examine national measures adopted in developed countries that can deter technology transfer, such as tariff peaks and tariff escalation that reduce technology flows and opportunities for learning and innovation in developing countries, and measures that limit access to results emanating from publicly funded R&D institutions and activities.[14] Moreover, since an effective technology transfer requires an active role by the recipient in adapting the technology to local circumstances and often innovating thereon, measures should be considered to encourage developing countries to engage in technological development.[15] Thus, subsidies implemented by developing countries with a view to strengthen their technology capacity should be treated as a non-actionable under WTO rules.

14) For instance, under US law preference should be given for the license of technology developed with federal funding to firms that agree to manufacture in the United States [35 USCA §204, 209(b)].

15) It is to be noted that funding for agriculture and rural development in poor countries has declined in the last two decades. For instance, the annual World Bank lending has dropped 47% over the past 12 years; the annual foreign aid by individual countries to agriculture fell by 57% between 1988 and 1996 ([18], p. 41).

12.5
Transferring Agro-Biotechnologies

A key issue in considering transfer of technology in biotechnology – as well as in other fields – is the extent to which technology transferred is suitable to the needs of the recipient country and can contribute to its sustainable development. The traditional model of technology transfer has failed, in particular, to reach the less advantaged producers in developing countries, such as the rural poor [19].

Some programs have been designed to suit the specific needs of developing countries. For instance, a number of initiatives have been set up in order to transfer new technologies to farmers in Africa, such as:

- The African Agricultural Technology Foundation (AATF)[16] – launched by the Rockefeller Foundation in March 2003 – aimed at making available new technologies developed by four major agricultural companies: Dow AgroSciences, DuPont, Monsanto and Syngenta.[17]
- The Global Partnership for Cassava Genetic Improvement, launched in November 2002 by the United Nations Food and Agriculture Organization (FAO) aimed at improving both yields and the nutritional value of cassava, the third most important source of calories in the tropics, including Africa.[18] The cassava partnership is made of 30 of the world's leading experts in cassava research, largely from public organizations such as the Brazilian Agricultural Research Corporation and the International Fund for Agricultural Development. There is also private sector involvement; The Danforth Center, a member of the cassava partnership, has been given a royalty-free license to use Monsanto's enabling technologies in the research.[19]
- Bt corn is being adapted to fight the corn-eating insects native to Egypt (northern Africa) and Indonesia. The project is one of several technology-sharing initiatives led by Michigan State University's Agricultural Biotechnology Support Program. Technology is being shared between DuPont – through its Pioneer Hi-Bred International, Inc., subsidiary – and the Egyptian Agricultural Genetic Engineering Research Institute.[20]
- Maize hybrids that are able to resist voracious stem borers are being developed by the International Center for the Improvement of Maize and Wheat, Kenyan Agricultural Research Institute, and the Syngenta Foundation for Sustainable Agriculture.[21]
- With the support of Dutch Ministry of International Cooperation, the Kenya/Netherlands Biotechnology programme was set up in 1993 under the auspices of the Kenya Agricultural Biotechnology Platform [now the Biotechnology Trust of Africa (BTA)]. The programme applied tissue culture to several crops, and

16) www.aatf-africa.org.

17) Biotechnology transfer: sharing innovation with the developing world: organizations work to give poor nations the tools and technology to improve their lives. http://www.whybiotech.com/index.asp?id=3650.

18) See Footnote 17.

19) See Footnote 17.

20) See Footnote 17.

21) See Footnote 17.

employed molecular marker technology for selection and maize breeding. The BTA is the largest biotechnology programme in Kenya, which has since mid-1990s supported eight biotechnology research projects on potato, cassava and sweet potato, banana, citrus, macadamia, biopesticide dust, marker-assisted breeding in maize, animal health, and institutional support [20].

- In the case of so-called *Golden Rice*,[22] Syngenta Seeds negotiated access to all technologies necessary for the production of the genetically modified rice [21] and granted the *Golden Rice* Humanitarian Board with the right to sublicense breeding institutions in developing countries free of charge, but for humanitarian use only. Syngenta is not interested in commercializing *Golden Rice* in developed countries since practically no vitamin A deficiency is found in such countries. The sublicenses are only for subsistence farming and not for commercial purpose (including exports).[23]

Although there is no attempt here to assess the costs and benefits of these projects for the recipient countries, it is worthwhile to note, first, that some of these projects focus on the adaptation and diffusion of products incorporating technologies, rather than on the transfer of technologies in an intangible form. The learning and capacity building impacts of these projects may, hence, be very limited. While technology may be transferred incorporated into equipment and other products which may help to increase production yields, improve quality or exploit new markets, they do not enhance the domestic technological capacity, which is crucial to promote sustainable development.

Second, as noted by Smale [22]:

> For crop biotechnology, as with any introduced technology, knowing the social determinants and social consequences of its adoption is important for designing policies to support its use. Although planting material may be neutral to the scale of the farm operation (that is, there is nothing inherent in the technology that implies large-scale farmers will have a greater ability to use it than smallholder farmers), a technology typically has an aspect that favors its adoption by certain social groups .... As compared with conventional genetic technologies, genetically transformed seed presents a number of unique challenges, such as the need to develop appropriate regulatory frameworks for biosafety.

Third, technologies transferred to developing countries and LDCs may displace local technological solutions and provide unsustainable modes of production. There is little doubt, for instance, about the advantages brought about by hybridization in some crops, such as maize. However, hybrid seeds cannot be replanted

22) *Golden Rice* has been genetically engineered to contain β-carotene and other carotenoids in the edible part of the grain. When the rice is consumed, some carotenoids are converted in the body into vitamin A. The use of this genetically modified plant provides a tool to address vitamin A deficiency in developing countries.

23) http://www.goldenrice.org/Content2-How/how9_IP.html.

and require farmers to buy new stocks of seeds every year. This is not only inconsistent with traditional practices of seed saving and re-use, but poses an economic burden that many poor farmers are not in a position to bear. New technologies, although valuable in themselves, may hence contribute little or nothing to poverty reduction if their introduction is not accompanied by other measures (such as financial support).

Fourth, the introduction of new technologies such as genetically modified organisms may – in addition to any other considerations regarding environmental or health impact – erect barriers for the export of production outputs. One illustrative case is provided by the detainment (on the basis of an expansive interpretation of European Regulation 1383/2003) of shipments of soymeal exported from Argentina by custom authorities of the Netherlands, Denmark and Spain, and the litigation in course in those countries against soymeal importers.

Soybean is currently the main culture of Argentina, with a participation of around 50% of the total seeded area of oily cereals and is one of its main export items (more than US$ 2 billion annually). Most soymeal is exported to Europe, which obtains around 50% of its soymeal consumption (mainly for animal feed) from Argentina.

Monsanto did not obtain a patent on its herbicide-resistant 'Roundup Ready' (RR) technology in Argentina, as it filed the respective application after the expiry of the applicable legal terms. The RR gene in soybeans was commercialized in 1996 and, thanks to the lack of patent protection, rapidly disseminated in the country. It is estimated that around 95% of soybean currently produced in Argentina is derived from varieties incorporating the RR gene. In fact, almost 200 varieties containing it have been developed since then (only a fraction by Monsanto itself), which are protected by Plant Variety Protection in Argentina.

The introduction of transgenic soybeans in Argentina without patent protection permitted Monsanto to rapidly disseminate them not only throughout Argentina, but also over Brazil and other South American countries. Sales of RR seed also boosted sales of Monsanto's glyphosate herbicide 'Roundup'.

In choosing to transfer its technology to Argentine seed producers, Monsanto voluntarily stimulated the production of transgenic soybean there. Although Monsanto obtained royalties for the RR technology under private contracts with seed companies, it aimed at getting an additional payment from Argentine farmers, who refused to pay an additional charge for a technology that is in the public domain. Monsanto targeted then the importation of Argentine soymeal into Europe, on the basis of two patents (EP 0218571 and EP 546090) that protect the gene and gene constructs as such, as well the transformed cells in a soybean plant. Thus, Monsanto attempts to use patents covering herbicide resistant genes to prevent trade in industrially processed products where such genes, if hypothetically still found intact, cannot perform its function.

On 9 August 2006, the DG Internal Market and Services of the European Commission provided an interpretation of Article 9 of the Directive on Biotechnological Inventions (98/44/EC).[24] It confirmed that derivative products, such as soymeal, are not covered by patent claims relating to generic information which do not perform their function in such products. Although it may be reasonably predicted that this paradigmatic case of 'strategic litigation' will end up with Monsanto's legal defeat, the resources invested by the Argentine government (which requested the status of affected third party in legal proceedings) and importers are very substantial.

This case illustrates a significant attempt to expand the legal powers conferred by patents covering genes. If these attempts were successful, they could have a major adverse effect on the transfer of materials to developing countries of materials that may contain patented genes in developed countries. Any derivative products (e.g. shirts made out of Bt cotton) would be potentially targeted by patent holders and imports encumbered or prevented in the developed countries where those genes are patented.

It should also be noted, that projects for the transfer of technology on a free basis to developing countries are subject in some cases to very specific conditions. For instance, under the *Golden Rice* initiative, sublicenses are only available for subsistence farming excluding any commercial purpose. This means that farmers adopting this technology will be prevented from exporting while the patents remain in force. It does not seem justifiable to introduce such a discrimination if the goal of initiatives of this type is to reduce poverty and contribute to the economic development of poor countries.

12.6 Transfer of Biotechnology for Industrial Production

So far, the use of modern biotechnologies for the production of industrial products, such as pharmaceuticals, is quite limited in developing countries. An interesting example of the transfer of technology for industrial applications (such as bio-pharmaceuticals) is provided by the activities of the International Center for Genetic Engineering and Biotechnology (ICGEB).

Most transfer of technology agreements entered into by the ICGEB (see Table 12.1) aim at the transfer of know-how for the production of recombinant proteins, including supply of the genetically engineered strain producing such proteins (mammalian cell, yeast or bacteria), supply of the protocols for production, purification and quality control, and training in ICGEB laboratories of a number of the partners' employees in the techniques required for successful completion of these activities.[25]

These transfers of technology might be particularly useful as 'over the next few years, the patents of a number of first-generation biopharmaceuticals will expire;

24) In accordance with Article 9 of the European Directive on Biotechnological Inventions (98/44/CE), the protection with regard to patents on a product containing or consisting of genetic information extends 'to all material, save as provided in Article 5(1), in which the product is incorporated and in which the genetic information is contained *and performs its function* [emphasis added]'.

25) In some cases, agreements may provide that ICGEB scientists assist its partners also after the completion of the transfer, until the pilot-scale level is reached.

Table 12.1 Agreements entered into by ICGEB for transfers of technology to the industrial sector.[26]

Product	Country	Year
Interferon-α2b	Sri Lanka	1999
	Iran	2001
	Italy	2002
	Venezuela	2003
	Pakistan	2003
	India	2006
Interferon-α2a	India	2002
	UAE (option)	2003
	South Africa	2003
	Brazil	2006
Interferon-β	Italy	1996
Cephalosporin C deacetylase	Italy	2001
CHO cell line	Sweden	2004
Development of a malaria vaccine	India	2001
Erythropoietin	Italy	1996
	Egypt	1999
	Sri Lanka	1999
	Italy/Iran	2001
	UAE	2003
	Venezuela	2003
	South Africa	2003
	Pakistan	2003
	Brazil	2005
	India	2006
Interferon-γ	India	1998
	Italy	1999
Granulocyte colony stimulating factor	Argentina	1995
	Italy	1999
	Sri Lanka	1999
	Cuba	2001
	UAE (option)	2003
	Egypt	2003
	Venezuela	2003
	South Africa	2003
	Pakistan	2003
	Turkey	2005
	India	2006
HCV antigen multi-epitope	India (non-exclusive)	2004

26) http://www.icgeb.trieste.it/SERVICES/BiotechTransfer.htm last accessed 11 December 2006.

Table 12.1 *(continued)*

Product	Country	Year
Hepatitis C diagnostic kit	Sri Lanka	1999
	India	2001
	United States	2003
Hepatitis B vaccine	Egypt	1999
	Iran	2003
HIV-1/2 diagnostic kit	India	1992
	Nigeria	1997
Human growth hormone	Italy	1996
Hybridoma clone	India	2003
Imiglucerase		
Interferon (pegylated)	Iran (non-exclusive)	2004
	Pakistan	2005
	Brazil	2005
Malaria compound	Italy	1997
Recombinant insulin	Argentina	1998
	Sri Lanka	1999
	Pakistan	2005
Thrombopoietin	Argentina	1995
	Italy	1996
Vegetable brassica expressing BT toxin	India	1999
Development of recombinant RHBV-BAC	India	1999
Development of CRY X-1 for cotton and CRY 1 LA5 and VIP for egg plant	India	2003
Development of a sprayable biopesticide	India	2003
Patent 'Molecular presenting system'	Licensed in United States	1998
Patent 'Screening methods ... p53'	Licensed in United States	1999
Patent 'Process for the production of interferon alfa'	Licensed in Italy	2002
Patent 'Vaccine'	Licensed in Argentina	2003
Patent 'Fine mapping and application of DNA markers to a gall midge resistance gene ...'	Licensed in India and United States	2003
Patent 'Chimeric polypeptides and their use'	Licensed in the European Patent Office (with exception of ICGEB MS, United States, Australia, Japan, Canada)	2003

this will open the market for the production of generic biopharmaceuticals with a potential global market of about US$ 2 billion'.[27]

12.7 Conclusions

In a growingly liberalized and globalized world economy, innovating firms may directly exploit their technologies on a world scale through trade: to the extent that tariff and other barriers have been reduced[28] or eliminated (in the context of an increasing number of bilateral and regional free trade agreements) such firms need not to look for licensing or foreign direct investment to enter a given market.

At the same time, efficient recipients of technology, may potentially compete and displace the firms that transferred the technologies they use, unless their commercial behavior is effectively controlled under contractual restrictive practices, such as those limiting exports. Hence, trade opportunities opened to both innovating firms and potential recipients may limit the incentives to transfer up-to-date technology.

If this hypothesis is correct, the implications for development are worrisome since, on the one hand, the possibility of imitation and reverse engineering was drastically curtailed by the TRIPS Agreement and, on the other, the capacity of developing countries to generate their own technologies, as mentioned, is limited. Developing countries may, in this scenario, only receive relatively mature technologies, which may not allow them to compete on the global market, or essentially remain as users of innovations generated in the developed world. As a result, transfer of technology should be regarded as *complementing* local innovation efforts: technology transfer, where available, needs to be used as a tool to further enhance local capabilities, rather than as the main engine of sustainable development.

This conclusion highlights the importance of current initiatives to avoid the proliferation of legal and technological measures that prevent access to scientific and other knowledge in the public domain.[29] This is particularly crucial in areas,

27) See Footnote 26.

28) The average tariffs amount to 5% in developed countries and to 28% in developing countries for non-agricultural products.

29) As stated in the 'Adelphi Charter':

> [I]n making decisions about intellectual property law, governments should adhere to these rules:
>
> - There must be an automatic presumption against creating new areas of intellectual property protection, extending existing privileges or extending the duration of rights.
> - The burden of proof in such cases must lie on the advocates of change.
> - Change must be allowed only if a rigorous analysis clearly demonstrates that it will promote people's basic rights and economic well-being.
> - Throughout, there should be wide public consultation and a comprehensive, objective and transparent assessment of public benefits and detriments.

Available at www.adelphicharter.org.

such as biotechnology, where the dividing line between 'science' and 'technology' is blurred and where the north–south asymmetries in technological capabilities are still very significant.

References

1 United Nations Industrial Development Organization (1996) *Manual on Technology Transfer Negotiation*, UNIDO, Vienna.

2 Dutfield, G. (2003) *Intellectual Property Rights and the Life Science Industries, A Twentieth Century History*, Ashgate, Dartmouth.

3 Correa, C. (1994) Trends in technology transfer – implications for developing countries. *Science and Public Policy*, 21, 369.

4 Byerlee, D. and Fischer, K. (2002) Accessing modern science: policy and institutional options for agricultural biotechnology in developing countries. *World Development*, **30**, 931–48.

5 Radosevic, S. (1999) *International Technology Transfer and Catch-up in Economic Development*, Edward Elgar, Cheltenham.

6 Kumar, N. (2002) *Intellectual Property Rights, Technology and economic Development: Experiences of Asian Countries*, Commission on Intellectual Property Rights, London.

7 Correa, C. (2005) From biotech innovation to the market: economic factors driving the South's competitiveness in biotechnology, in *Trading in Genes: Development Perspectives on Biotechnology Trade and Sustainability* (eds R. Meléndez-Ortiz and V. Sánchez), Earthscan, London.

8 Correa, C. (2005) Can the TRIPS agreement foster technology transfer to developing countries?, in *International Public Goods and Transfer of Technology Under a Globalized Intellectual Property Regime* (eds K. E. Maskus and J. H. Reichman), Cambridge University Press, Cambridge.

9 Maskus, K. (2005) The role of intellectual property rights in encouraging foreign direct investment and technology transfer, in *Intellectual Property and Development. Lessons from Recent Economic Research* (eds C. Fink and K. Maskus), World Bank/Oxford University Press, Washington, DC.

10 Correa, C. (2007) *Trade Related Aspects of Intellectual Property Rights. A Commentary on the TRIPS Agreement*, Oxford University Press, Oxford.

11 Reichman, J. (2006) *Nonvoluntary Licensing of Patented Inventions: The Law and Practice of the United States*, UNCTAD-ICTSD, Geneva.

12 Oh, C. (2006) Compulsory licences: recent experiences in developing countries. *International Journal of Intellectual Property Management*, **1**, 22–36.

13 Oddi, S. (1996) TRIPS – natural rights and a 'polite form of economic imperialism'? *Vanderbilt Journal of Transnational Law*, **29**, 415–70.

14 Throsteinsdsttir, H., Quach, U., Daar, A. and Singer, P. A. (2004) Health biotechnology innovation in developing countries. *Nature Biotechnology*, **22** (Suppl), DC48–52.

15 Jullien, E. (1989) *Les impacts économiques de la protection de l'innovation sur le secteur européen de la sémence*, CERNA, Paris.

16 Xu, B. and Chiang, E. (2005) Trade, patents and international technology diffusion. *Journal of International Trade and Economic Development*, **14**, 115–35.

17 Mytelka, L. (2002) Development: IPRs, TRIPS hamper R&D, technology transfer to South, submission to the WTO Working Group for the Study of the Inter-relationship between Trade and Transfer of Technology, SUNS #5139.

18 Paarlberg, R. (2001) Environmentally Sustainable Agriculture in the 21st Century, in *Aspen Institute Congressional Program*, Aspen Institute, Queenstown, MD.

19 Nyangito, N. and Okello, J. (1998) *Kenya's Agricultural Policy and Sector Performance: 1964 to 1996*, Institute of Policy Analysis and Research, Nairobi, available at http://www.ipar.or.ke/OP-04.htm.

20 Odame, H., Kameri-Mbote, P. and Wafula, D. (2003) *The Role of Innovation in Policy and Institutional Change: The Case of Transgenic Sweetpotato in Kenya*, International Environmental Law Research Centre, Geneva, available at http://www.ielrc.org/content/n0206.htm.

21 Kryder, D., Kowalsi, S. P. and Krattiger, A. F. (2000) *The Intellectual and Technical Property Components of Pro-Vitamin A Rice (GoldenRice™): A Preliminary Freedom-To-Operate Review, ISAAA Briefs 20*. ISAAA, Ithaca, NY.

22 Smale, M. (2006) *Assessing the Impact of Banana Biotechnology in Kenya. ISAAA Briefs 19*. ISAAA, Ithaca, NY.

Index

Technology Transfer in Biotechnology. A Global Perspective.
Edited by Prabuddha Ganguli, Rita Khanna, and Ben Prickril

ISBN: 978-3-527-31645-8